Bicycle Enginee and Technolog'

Bicycle Engineering and Technology is a primer and technical introduction for anyone interested in bicycles, bicycling and the bicycle industry. With insight into how bicycles are made and operated, the book covers the engineering materials used for their manufacture and the technicalities of riding. It also discusses ways in which the enthusiast may wish to get involved in the business of working with these fantastic machines, which are now being aided with electrical power.

The bicycle is a significant factor in transportation around the world and is playing an increasingly crucial role in transport policy as we collectively become more environmentally conscious. To celebrate the importance of the bicycle on the world stage, a brief history is included along with a detailed timeline showing the development of the bicycle with major world events.

Previous knowledge of engineering or technology is not required to enjoy this text, as all technical terms are explained and a full glossary and lists of abbreviations are included. Whether you are a bicycling enthusiast, racer, student or bicycle professional, you will surely want to read it and keep it on your shelf as a handy reference.

Andrew Livesey MA CEng is a lecturer in Engineering at Ashford College University Centre in Kent, England, when he is not riding his bicycle or building something in his garage. He is a member of London Clarion CC and Thanet RC, mixing social riding with charity events and club competitions—he describes himself as an all-round clubman. He also enjoys very fast motorcycles and high-performance cars. He has previously published with Routledge *Basic Motorsport Engineering* (2011), *Advanced Motorsport Engineering* (2012), *The Repair of Vehicle Bodies, 7th Edition* (2018) and *Practical Motorsport Engineering* (2018).

Bicycle Engineering and Technology

Andrew Livesey

Routledge
Taylor & Francis Group

LONDON AND NEW YORK

First published 2021
By Routledge
2 Park Square, Milton Park, Abingdon, Oxon OX14 4RN

and by Routledge
52 Vanderbilt Avenue, New York, NY 10017

Routledge is an imprint of the Taylor & Francis Group, an informa business.

British Library Cataloguing-in-Publication Data
A catalogue record for this book is available from the British Library

Library of Congress Cataloging-in-Publication Data
Names: Livesey, Andrew., author.
Title: Bicycle engineering and technology / Andrew Livesey.
Description: Abingdon, Oxon ; New York, NY : Routledge, 2021. | Includes index.
Identifiers: LCCN 2020031641 (print) | LCCN 2020031642 (ebook) | ISBN 9780367419172 (hardback) | ISBN 9780367419165 (paperback) | ISBN 9780367816841 (ebook) | ISBN 9781000259261 (epub) | ISBN 9781000259124 (mobi) | ISBN 9781000258981 (adobe pdf)
Subjects: LCSH: Bicycles--Design and construction. | Bicycles--History.
Classification: LCC TL410 .L58 2021 (print) | LCC TL410 (ebook) | DDC 629.227/2--dc23
LC record available at https://lccn.loc.gov/2020031641
LC ebook record available at https://lccn.loc.gov/2020031642

ISBN: 978-0-367-41917-2 (hbk)
ISBN: 978-0-367-41916-5 (pbk)
ISBN: 978-0-367-81684-1 (ebk)

Typeset in Sabon
by KnowledgeWorks Global Ltd.

Visit the eResources: www.routledge.com/9780367419172

Dedication

This book is dedicated to all those cyclists who
have helped me on my life-time long cycling journey,
and the clubs to which they belong. The clubs, old
and new, in no particular order, include:

Burnley Clarion CC, Clayton Velo, London Clarion CC,
North Lancashire Road Club, Thanet Road Club, V C Meudon.

Special individual thanks to: Edward Gilder, Ian Clarke,
Alex Southern, Tony Hadland, Mike Burrows, James Burrows
and Jay de Villiers and perhaps most importantly to the
late Adam Hill for the creation of Hill Special Cycles, he
is known as the father of the lightweight cycle industry
and his creations the Rolls Royces on two wheels.

The author on his Hill Special, Thanet Road Club
Hill Climb Championships 2019, Kent.

Additional information on Health and Safety Risk
Assessments, Data Sheets, Apprentice Standards and
gear ratio calculations can be found on the associated
website: www.routledge.com/9780367419172

Contents

Foreword

The safety bicycle, the forerunner of today's modern machine is now well over a hundred years old. This apparently simple machine transformed the lives of millions, both economically and socially, stretching horizons from places no more than a day's walk away, to places far and wide. It provided unparalleled access to more distant places, both for work and leisure. Places which previously could only have been dreamt of. Unbelievable distances were achievable if you just kept on pedalling.

Many words have been written since that time on cycling as a sport, as a pastime, as an engineering challenge or just as every day transport. But there has been no recent volume which has encompassed all of these until now. Andrew, with his lifelong passion for the sport, years of membership of the National Clarion Cycling Club and a background in motorsport engineering has once again combined all that we know, and much that we don't about the bicycle, in to a single volume. Bang up to date—here we have history, technology, practicality and joy, delivered with skill and clarity for a 21st century audience.

This is a volume that cyclists will want to return to time and again, with its analysis of cycling over time, how the bicycle has developed as transport and at the other extreme, how it has become a gladiatorial machine for sportsmen and women, both on and off the road and on the track.

When Tom Groom called the first meeting of the Clarion Cycling Club in Birmingham in 1894, I am sure that he and his fellow members didn't realize just how important the bicycle would become in the social and economic history, not only of this country but the world. The Clarion slogan "Fellowship is Life" remains with us to this day along with the National Clarion Cycling Club and its more local sections. They, along with many other varied national and international organizations continue to promote the bicycle as a machine for the 21st century. One ever more relevant in these times of climate change. Once a rider, racing, on the road to work, out for a leisure ride with your family or even on a round the world tour, you will always find fellowship with others when you are on your bicycle.

Edward Gilder
Editor Boots and Spurs
The magazine of the National Clarion Cycling Club

Preface

Bicycles have developed over the past two centuries, the way we use them as developed too. They have become much more accessible, most households in the UK have more than one bicycle and there is an increasing use of electric bicycles. The sport and past-time of bicycling is accessible to just about everybody, giving the benefits of good health and reduced pollution.

If you are new to bicycling, I would recommend you to join your local cycling club to enable you to access the local network of cyclists and informal events, most clubs hold one or two events each week, and affiliate to one of the national bodies which offer insurance to cover should you be involved in an accident.

Andrew Livesey MA CEng
Herne Bay, Kent
Andrew@Livesey.US

Abbreviations and symbols

The abbreviations are generally defined by being written in full when the relevant technical term is first used in the book. In a very small number of cases, an abbreviation may be used for two separate purposes, usually because the general concept is the same, but the use of a superscript or subscript would be unnecessarily cumbersome, in these cases, the definition should be clear from the context of the abbreviation. The units used are those of the internationally accepted *System International* (SI). However, because of the large American participation in cycling, and the desire to retain the well-known Imperial system of units by UK cycling enthusiasts, where appropriate Imperial equivalents of SI units are given. Therefore, the following is intended to be useful for reference only and is neither exhaustive nor definitive.

α	(alpha) angle—tyre slip angle
λ	(lambda) angle of inclination
μ	(mu) co-efficient of friction
ω	(omega) rotational velocity
ρ	(rho) air density
η	(eta) efficiency
θ	(theta) angle
a	acceleration
A	area—frontal area of bicycle and rider; or Ampere
ABS	anti-lock braking system; or acrylonitrile butadiene styrene (a plastic)
AC	alternating current
AF	across flats—bolt head size
AFFF	aqueous film forming foam (fire-fighting)
ATB	All Terrain Bike, synonymous with MTB
bar	atmospheric pressure—101.3 kPa or 14.7 psi as standard or normal
BATNEEC	best available technique not enabling excessive cost
BS	British Standard
BSI	British Standards Institute

C	Celsius; or Centigrade
CAD	computer-aided design
CAE	computer-aided engineering
CAM	computer aided manufacturing
C_D	aerodynamic co-efficient of drag
CG	centre of gravity, also CoG
CIM	computer integrated manufacturing
C_L	aerodynamic co-efficient of lift
cm	centimetre
cm^3	cubic centimeters—capacity; also called cc. 1000 cc is 1 litre
CO	carbon monoxide
CO_2	carbon dioxide
COSHH	Control of Substances Hazardous to Health (Regulations)
CP	centre of pressure
CR	compression ratio
D	diameter
d	distance
dB	decibel (noise measurement)
DC	direct current
deg	degree (angle or temperature), also^0
dia.	Diameter
DTI	dial test indicator
EC	European Community
ECU	electronic control unit
EPA	Environmental Protection Act; or Environmental Protection Agency
EU	European Union
f	frequency
F	Fahrenheit, force
ft	foot
ft/min	feet per minute
FWD	front-wheel drive
g	gravity; or gram
gal	gallon (USA gallon is 0.8 of UK gallon)
GRP	glass reinforced plastic
HASAWA	Health and Safety at Work Act
HGV	heavy goods vehicle (used also to mean LGV - large goods vehicle)
hp	horse power (CV in French, PS in German)
HPV	Human Powered Vehicle
HSE	Health and Safety Executive; also, health, safety and environment
I	inertia
ID	internal diameter

IMechE	Institution of Mechanical Engineers
IMI	Institute of the Motor Industry
in^3	cubic inches—measure of capacity
IR	infra-red
ISO	International Standards Organization
k	radius of gyration
kph	kilometers per hour
l	length
L	wheelbase
LH	left hand
LHD	left hand drive
LHThd	left hand thread
LPG	liquid petroleum gas
lumen	light energy radiated per second per unit solid angle by a uniform pointsource of 1 candela intensity
lux	unit of illumination equal to 1 lumen/m^2
M	mass
MAX	maximum
MIG	metal inert gas (welding)
MIN	minimum
MTB	Mountain bike, also known as ATB
N	Newton; or normal force
Nm	Newton metre (torque)
No	number
OD	outside diameter
OL	overall length
OW	overall width
P	power, pressure or effort
Part no	part number
PPE	Personal Protective Equipment
pt	pint (UK 20 fluid ounces, USA 16 fluid ounces)
PVA	polyvinyl acetate
PVC	polyvinyl chloride
Q	heat energy
r	radius
R	reaction
Ref	reference
RH	right hand
rpm	revolutions per minute; also RPM and rev/min
RTA	Road Traffic Act
RWD	rear-wheel drive
std	standard
STP	standard temperature and pressure
TE	tractive effort
TIG	tungsten inert gas (welding)

V	velocity; or volt
VOC	volatile organic compounds
W	weight
w	width
WB	wheel base
x	longitudinal axis of vehicle or forward direction
y	lateral direction (out of right side of vehicle)
z	vertical direction relative to vehicle

Superscripts and subscripts are used to differentiate specific concepts.

SI UNITS

cm	centimetre
K	Kelvin (absolute temperature)
kg	kilogram (approx. 2.25 lb)
km	kilometre (approx. 0.625 mile or 1 mile is approx. 1.6 km)
kPa	kilopascal (100 kPa is approx. 15 psi, that is atmospheric pressure of 1 bar)
kV	kilovolt
kW	kilowatt
l	litre (approx. 1.7 pint)
l/100 km	litres per 100 kilometres (fuel consumption)
m	metre (approx. 39 inches)
mg	milligram
ml	millilitre
mm	millimetre (1 inch is approx. 25 mm)
N	Newton (unit of force)
Pa	Pascal
ug	microgramImperial Units
ft	foot (= 12 inches)
hp	horse power (33,000 ftlb/minute; approx. 746 Watt)
in	inch (approx. 25 mm)
lb/in^2	pressure, sometimes written psi
lbft	torque(10 lbft is approx. 13.5 Nm)

Chapter 1

Brief history of bicycles and bicycling

Hobby Horse to balance bike—a brief history of bicycles and bicycling

A wise person will always do research, that is, looking what others have done in a particular field before starting on a new project. Although, to the untrained eye, cycles are seen as simple mechanical machines within the field of engineering, they cut across many other academic fields and, therefore, both the bicycle and bicycling history are tied up with the history and development in other fields.

The purpose of this chapter is to give you an insight into the history and heritage of bicycles and bicycling to help put this activity of ours into the context of the wider development going on in the world.

In all the fields of engineering and technology, there is always more than one person or organization working on design and development in any area. They tend to work alone, or in small groups to keep a

Figure 1.1 Draisienne or Hobby Horse.

Figure 1.2 **Chinese wheelbarrow.**

technological and financial advantage over others. Sometimes new technology is not made available until after the death of the inventor. UK Government research is often kept under the Official Secrets Act for 30 years before being made available to the wider public. Because of this situation, it is often not clear who invented or developed any particular item or feature first.

The historical timeline is an attempt to place the development of bicycles and bicycling in both a chronological order, and in relationship to other events going on in the world.

A good historical example of the transfer of skills and knowledge from one field to another is illustrated by the company Lines Bros. Ltd. The two brothers are perhaps better known by their range of children's cycles called Tri-ang, now defunct, but still often seen in use. The Lines Brothers inherited a toy business from their uncle, one of the brothers studied woodwork at evening classes and later architectural design. The business grew producing a large range of toys, mainly from pressed steel. When World War II broke out, the British troops used American Thompson sub-machine guns, The Thompson factory was unable to meet the volume demanded by the British Army. The Lines Brothers stepped in, redesigning a British machine gun, so that it could be made in their toy factory using the machine tools and assembly lines which had been used for toy making. They made over a million STEN machine guns and 14,000,000 parts for Spitfires. After the war, this production was converted to making children's bicycles.

The development of the bicycle is also interlinked with the development of cars and motorcycles; but perhaps more importantly with the social and political changes in the world, and the development of materials and engineering production techniques.

NEEDS AND WANTS

When we talk about historical events and technological developments, we should remember that these are always developed to either satisfy a need or because somebody had a particular desire to do something special. When we look at cycle developments, this is exactly the case. It is also worth noting that most advances in science and engineering technology have been made either by upper-class people with time and money to investigate their interests, or hard-working poor people trying to resolve a social situation or a particular achievement.

Bicycles would never have been invented and developed if there wasn't a need or want for one. Let's have a look at what happened and how it happened. Let's go back to about the year 1700. London had started to recover from the bubonic plague, or black death, caused by rats. It killed 70,000 people in London and about half the population in India too. The great fire of London had destroyed the old City of London, doing a job of slum clearance and getting rid of the rats. So, in 1700, re-building work was underway in London and life was changing. The Bank of England was newly established and insurance was now available through Lloyds of London. Similar things were happening in the rest of the world. It was the start of the Georgian period—the Kings of England were all called George for the next 130 years.

At this point nobody wanted a bicycle, there was no need for one. However, things were changing. The next 100 years saw the American War of Independence, a revolution for self-rule and the French Revolution. Both with lots of bloodshed, but the most important revolution was about to follow—the Industrial Revolution. The building of factories and steam powered machinery changed how people lived and worked. This led to a need for transport, especially internal transport, for moving goods and people around the growing towns and between the towns.

EARLY BICYCLES

Let's have a look at how the bicycle developed over what was really a very short period of time. The industrial revolution had kicked in across the whole developed world at the start of the 19th Century. The engineering technology existed to build most things out of metal, and it was becoming fairly common place. Workshops were being set-up to make machines and tools. Engineering was becoming a profession with more people studying mathematics and mechanics both at university and in the growing number of municipal mechanics institutes in the provincial towns.

Tech note

Mechanics in this sense refers to the physical sciences, it is the correct name for the amalgam of pure physics and engineering science. Mechanics institutes, these are what have now become technical colleges, ran evening and weekend courses for mainly men who wanted to progress in the engineering and manufacturing trades and professions. They learnt the mathematics and mechanics of machines and how to make and repair them.

Draisienne or Hobby Horse

This is accepted as probably being the first bicycle. How it actually came about is subject to conjecture and propositions, but the story which is told seems very plausible. Baron Karl von Drais, who lived in the town of Mannheim in what is now Germany, studied mathematics, mechanics and architecture at Heidelberg University. It is worth noting that Heidelberg is the oldest university in Germany. Heidelberg University was the setting for the operetta *The Student Prince,* very much the place to be for young clever people, so we can suspect that Baron Karl von Drais was of such a class, and interestingly Mannheim is the town where the Diesel engine was invented. The two towns are about 15 miles apart—the journey would take a little over an hour in a carriage then, 15 minutes now. There was a gigantic volcanic eruption in 1816, so much so that it became known as the year without a summer all across central Europe. This sent the prices of food up, meaning that oats for horses were very expensive and people were looking for other forms of transport. Hence the invention of the Draisienne. When I read about this the first time, I had a problem of actually picturing the use of a Draisienne. I couldn't see it on the cobbled streets of Burnley. However, now I live in a sea-side town in Kent, I see all sorts of contraptions being ridden on the smooth surface of the promenade in front of my house. A recent craze has been the balance bike for young children—this is in effect a mini Draisienne.

The Draisienne grew in popularity, and were built and copied in many countries including the UK and America. In the UK, the name Hobby Horse was used—probably to hide the fact that it had not been invented in the UK, and avoid patent royalties. The Draisienne had 27-inch wheels (690 mm), the same as current bicycles, probably to allow the rider's feet to touch the ground when seated on the saddle. The Draisienne also had a steering mechanism, like modern bicycles, and you needed to be able to steer the front wheel not just to go around corners, but also to maintain balance.

It is worth comparing the shape and mechanics of the Draisienne with the Chinese wheelbarrow. The Chinese wheelbarrow was one of the items that Baron Karl von Drais would have studied in mechanics lessons at university. It is about the load being directly above the wheel, so that when evenly loaded, only a light force is needed to raise the rear legs off the ground, allowing the muscles to concentrate on pushing it forwards.

Of course, the Draisienne didn't have any drive mechanism. I can imagine Baron trying to push, shove and coast the bicycle between the university in Heidelberg and his home Mannheim, especially after a few beers' student prince style, it's a fairly straight road. Baron moved on to invent other things and become a teacher. So, there are now quite a lot of Draisienne/Hobby Horse bicycles around the world made from both metal and wood, these are working their way down in value as they age. So, obviously other people are going to think how they can apply pedals or other ways to make them go. Strangely this problem, or at least curiosity, still exists with the current effort going into electric powered bikes, as well as electronic gears and simple shaft drive.

Figure 1.3 **Boneshaker.**

Figure 1.4 Handle bars on boneshaker.

Figure 1.5 Rear brake on boneshaker.

Figure 1.6 Detail on boneshaker.

Figure 1.7 Saddle on boneshaker.

Figure 1.8 Boneshaker clamped.

Figure 1.9 Boneshaker in use in London on Tweed Run.

Figure 1.10 Pedal mechanism on boneshaker.

Around 1839, Kirkpatrick Macmillan, aged about 27 years old, fitted a sort of crank and lever mechanism to enable the back wheel to be driven by the pedal action. The bicycle is now named the velocipede—this is a non-pure Latin translation of swift-foot. Macmillan is described in the popular books as a blacksmith, he may indeed have worked as one with his father; but I suspect that he was well read in mathematics and mechanics. With the amount of iron and steel, the velocipede would be very heavy and, therefore, hard to pedal up any hills, of which there are many in Scotland where Macmillan lived. Lots of other people produce variations of bicycles, tri-cycles and indeed quad-cycles. The search is on for a lighter bicycle, one which is truly rideable, unlike the others which are only used for promenading along the seafront Brighton, or the likes of Hyde Park in London. Maybe Central Park in New York, or at the side of the Seine in Paris.

Ordinary or Penny-farthing

In about 1870, James Starley invented the bicycle that we now know as the Penny-farthing. James Starley was first employed as a sewing machine engineer in Coventry, before going into bicycle manufacturing. The Penny-farthing was the bicycle that was needed to make bicycling available to the wider population. They were cheap, light and ridable. These are the criteria by which we still judge bicycles. It is worth analysing these points to show the technical development:

- The Penny-farthing uses a fairly simple one-piece tubular frame, much light and cheaper than the complex Draisienne assembly.

- The wheels are steel spoked, lighter than the wooden artillery wheels and therefore, easier to rotate.
- The crank mechanism is conveniently placed and easy to use.
- The wheel diameter is between 56 inches and 69 inches, this gives a good gear ratio to pedal turns—in fact, this is the gear ratio of current single-speed bicycles—making it very rideable.

James Starley also made a number of other bicycles, tri-cycles and quad-cycles. He also made Coventry the home of bicycle manufacturing, a fact that is still celebrated by the city.

In 1870, the world was a fairly calm and prosperous place; we have had the French Revolution, the American Civil War was over and the famine of the 1840s had passed. The world was building railways and steam ships. Perhaps for cycling, an important milestone was the granting of the Saturday half-holiday. The skilled workers, later followed by others, had the working week reduced. What better way to enjoy the weekend than by bicycling? The skilled classes could now afford one of these new Penny-farthings and they had the time to ride them too.

As the 19th Century turned into the 20th Century, the price of bicycles had gone down. The rich aristocrats and others with lots of money were now spending their time and money on motor cycles, cars and even airplanes.

Cycling as a hobby, past-time and sport started to become popular and affordable. The Cyclist Touring Club was founded in 1878 and other clubs followed. This was, however, a man's sport and women wanted to join in. The Penny-farthing was still a challenge to ride. James Starley's nephew John Kemp Starley designed the Rover Safety, a bicycle very similar in appearance to the current roadster/tourer type of bicycle. It was fitted with 26 inch wheels; this wheel size is still used on this type of bicycle. The smaller wheels were possible because of the use of chain and sprocket gearing. With the smaller wheels, step-through style frame and mudguards, ladies could ride a bicycle in comfort.

Another manufacturer, Raleigh Cycles was founded in 1887, in Nottingham. They opened a five-storey factory and employed over 200 staff. They introduced The All Steel bicycle which became an instant best seller, a modern version of it is still sold today. The Rover Company moved on to making motorcycles, then a range of cars and, of course, the well-known Land Rover vehicles. There were many other cycle manufacturers at THIS time, many of whom went on to become car manufacturers.

When the war broke out in 1914, the bicycle industry reacted in two ways. One was to make tens of thousands of bicycles for the Army Cyclist Corps. In fact, at one point, all the regiments had platoons of cyclists; they had replaced horses for carrying urgent messages. The other work that the bicycle manufactures did was to convert their workshops to making munitions—that is weapons, ammunition, equipment and stores.

As cycling was at this point a very popular sport and past-time for skilled workers, there was no shortage of recruits into the Army Cyclists Corps at the start of the war. Army cycling had actually started in 1908 with the Territorial Cyclist Battalions—part-time soldiers. Cyclists in the Army Cyclists Corps were payed higher than the same ranks in other jobs, providing that they met certain levels of ability.

Figure 1.11 WWI folding bicycle.

Figure 1.12 Hetchins curly rear stays, c. 1935.

Figure 1.13 Grocers' delivery bicycle, 1950.

Following the end of World War I, the costs of cars and motorcycles went down. In 1922, there were over 2,000 individual car companies in the UK vying for business. So, the usage of bicycles declined. The development of what became known as the lightweight bicycle in the 1930s gave the industry a boost for a while, to be cut short by World War II. Also, the great depression from the mid-1920s to World War II hit individuals and firms all across the Northern Hemisphere. After World War II, there was again an interest in cycling, but this was short lived until about 1957 when cars became almost as cheap as motorcycles. The fashionable chopper bicycles, BMX and then mountain bikes kept sales going, but lightweight sales almost died until the popularity of cycling in the 2012 Olympics.

MIKE BURROWS

Probably the biggest changes in the concept of bicycles, that is, how we look at bicycles and perceive the design of them, has been brought about by rider and designer Mike Burrows. Time trialling is about covering a set distance in the minimum time, unlike road racing which is solely about getting over the finish line first. So, for time trialing, outright speed is the name of the game, hence a desire to design a bicycle to go as quickly as possible with any given rider.

Mike Burrows' research and design are of a very practical nature—analysing current bicycle design, applying engineering knowledge and rider experience to the analysis and coming up with a new design, then seeing if it actually works in practice. He has been very successful and influential in the development of bicycles, particularly in the area of aerodynamics. Given a smooth and flat road or track, a strong rider and the correct gearing, the next biggest issue to then resolve is aerodynamics. Three big contributions to the field of aerodynamics made by Mike Burrows are discussed next.

Monoblade front fork

This came about after Mike Burrows saw one made in about 1889 by the Surrey Machinists' Company. Interestingly, the bicycle which it is used on also has an interesting cross-shaped frame. The design and the use of a monoblade fork on a modern bicycle are much more of a challenge than on a bicycle of the 19th Century, this is because of the weight restrictions and the need for greater precision. The use of the monoblade cuts the air resistance of the fork assembly in half and, similarly, reduces the weight.

Figure 1.14 Monofork on electric bicycle.

Monocoque frame

The design of the monocoque to replace the diamond frame apparently came from Mike Burrows seeing the sculptures of Barbara Hepworth—these abstract shapes have flowing lines of an almost organic nature.

Faired human-powered vehicles

Mike Burrows has produced a large number of human-powered vehicles (HPVs) over the years and taken the design of HPV to the highest level.

Some of Mike Burrows' designs are shown next.

Photographs taken by **MIKE BURROWS AND PAUL BURROWS** of bicycles designed by Mike Burrows. Thank you to both Mike and Paul, and to Tony Hadland for facilitating this through Hadland's Blog.

Figure 1.15 Last monocoque—designed for, but never ridden by, Abraham Olano in 1999. Abraham Olano is the only racing cyclist to have won both the World Road Race Championship and the World Time Trial Championship. The monocoque is made from pre-preg carbon fibre in two halves bonded together. The aerodynamic chain case houses a chain which is especially made with links almost half the size as those on other bicycles.

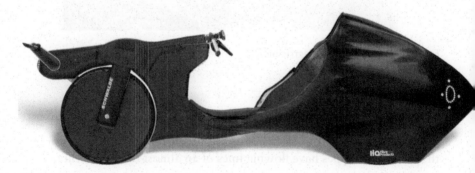

Figure 1.16 Rat racer excession—a part-faired HPV built for racing in 2014; it won a number of races. Its style is similar to that of formula racing cars, very purposeful shape.

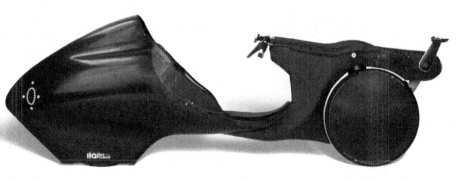

Figure 1.17 Daisy—a fully enclosed HPV. Built in conjunction with other designers and the TV bicycle and motorcycle racer, Guy Martin. The construction is a mixture of welded aluminium and carbon fibre sandwiched with a honeycomb filler.

Figure 1.18 Rat racer GT—a minimalist, back to basics design approach in 2015 for a recumbent bicycle used for please rides around the English country lanes.

Figure 1.19 Gordon—designed in 2015, and now becoming the basis for many commuter bicycles and electric bicycles. It is reminiscent of 1960s mopeds and old Dutch step throughs. A style seen on London streets as Santander bicycles. This one is amazingly light weight carbon fibre with a pinion gear drive.

CYCLING CLUBS

Cycling is a mixture of an individual sport and a team activity. Like all human-driven activities, it has always been competitive. There has always been the challenge of riding faster and further than the other rider. It isn't easy to train and beat others, but at the same time share your passion with them. Of course, you need others for time keeping, marking out the course and organising the event. So, cyclists have always tended to form groups or cycling clubs. Some clubs concentrate on racing, while some on simply riding together for comradery.

Figure 1.20 Classic bicycles at an event.

Most large towns have a cycling club, some have several. Each club has its own history. The clubs are mostly too small to be able to financially provide the services needed by individual cyclists; these services include insurance, event organization, dealing with transport issues and other macro-level activities. So, at a level above the local clubs, are regional and national bodies working to make cycling the sport that it is. This next section briefly looks at the history of the national cycling bodies in the UK.

> **British Cycling:** Founded in 1959—formerly called the British Cycling Federation (BCF)—it came about with the very fiery amalgamation of the National Cyclists' Union and the League of Racing Cyclists. Literally, members of each of the two formative organizations fought on the road. The National Cyclists' Union wanted to limit cycle racing to cycle tracks, while the League wanted to race on the roads.

This led to strong political fights as well as the physical fighting. The BCF embraced road events, but the League members saw it as loosing, so there are still branches of the League organising events on roads. BCF is based at the Manchester Velodrome, they are involved in both track and road racing and organise The Great Britain Cycling Teams for international events and the Olympic Games.

Cycling Time Trials: Founded in 1937—formerly called the Road Time Trials Council. Bicycle racing on roads was illegal in the 1930s. Indeed, it was illegal until the formation of the BCF in 1959. However, riders used to meet on quiet roads in the early hours of Sunday morning to race against the clock. They would be dressed in black alpaca suits and the riders' names were taped over to hide their identity. They were like Ninja warriors, all looking alike to hide their identity. The courses were given code numbers to prevent them being identified; these course codes are still in use today as they are part of what makes time-trialling such a popular sport. Most cycling clubs are also affiliated to Cycling Time Trials because of the wide range of time-trials, especially club mid-week events.

Cycling UK: Founded in 1878—formerly known as the Cyclist Touring Club (CTC). Cycling UK has what are called sections, i.e., clubs within a club, in most towns. They tend to have retained their names and identity, for example Blackburn and District CTC. To reflect the diversity of interests in cycle clubs, touring and racing, many of the CTC sections are also affiliated to British Cycling.

National Clarion Cycling Club: Founded in 1895—the Clarion was originally founded for the promotion of socialism, taking its name from

Figure 1.21 Clarion Cycling Club outing.

the newspaper with the same name. The Clarion members started by riding to deliver socialist—Labour Party—journals. This was before the formation of the Labour Party as we now know it. They had a large gathering over the Easter Weekend of 1895, known as The Easter Meet. Eventually, the membership spread out across the country into what are called sections, like those of the CTC. The political calling continues; but at a very low level, Easter Meets continued to be held in different parts of the UK each year. The Clarion Cycling Club is part of a much larger informal group of Clarion Clubs which include choirs, working Men's' Clubs and walking groups; but they are not formally joined, only by informal association. The Clarion between about 1900 and 1930 set up a number of country cottages and tea-rooms for members to visit and to get working people away from the Smokey towns. A few of these remain run by volunteers.

Saint Christopher's Catholic Cycling Club: Founded in 1932—a club set up to allow like-religiously minded people to enjoy cycling; they have regular meets, holiday tours and occasional pilgrimages. The club is recognised by the Vatican and the Archbishop of Westminster.

Union Cycliste Internationale (UCI): Founded in 1965—a body which came into existence at the same time as road racing became legal in

Figure 1.22 Sustrans signpost.

the UK; leading to the recognition of professional cycle racing in the UK and now across the world. The UCI set internationally accepted rules for racing cyclists, racing bicycles and events. British Cycling, because of its successes in World Tour events, plays a major role in the governance of the UCI.

David Gordon Wilson and Jim Papadopoulos—what they brought to bicycle design and bicycling, in general, was a serious scientific approach to the subject. They contributed over a century of research and questioning about how bicycles work. This was coupled with a love for the sport and past-time of bicycling.

Figure 1.23 Safety bicycle with wooden rims.

Figure 1.24 Penny-farthing in use in London.

Figure 1.25 Step for getting on Penny-farthing.

Figure 1.26 Using a bicycle for collecting cardboard packaging at Wuhan University of Technology in China, where the author worked for a while.

TIMELINE

This timeline is a guide to enable the reader to see the developments of the bicycle along with other developments in engineering, science and technology and in relationship with other major socio-political events. History is mostly unclear, even if we were there on the day, we still may not have seen the actual things that triggered an invention or other change, however viewing it in context helps to define the moment and may help to provide an insight for future bicycle engineers. All dates are approximate, the timeline reflects commonly accepted data.

Date	Bicycle development	Science, technology and arts	Other socio-political events
1600– 1800	Goods and passenger carrying coaches were developed to a high level with regular services between major towns. Springs, bearings and metal tyres are in use. These horse-drawn vehicles had efficient brakes and steering systems. So, the basic materials and manufacturing capability for bicycles existed.	The start of the new century—1600 marked the beginning of the scientific revolution—Galileo and Bacon started examining nature; they used *scientific method* as we use today.	Re-organization of postal services. Sunday Observance Act was enforced. First Turnpike Act, charging road users; Bubonic plague; fire of London; Lloyds insurance started; and Bank of England was founded.
	What didn't exist was the understanding of balance for bicycles until this was found with the *Hobby Horse* in the safe environment of a public park, although one can imagine boys playing *hoop and stick* with old cart metal tyres.		1760–1830: Industrial Revolution. 1776: American Independence. 1789–1799: French Revolution.
1818	Draisienne/*Hobby Horse* bicycle of Baron von Drais de Sauerbrun	1823: Introduction of Rugby.	1830–1890: Universities started higher education for all.
1839	Kirkpatrick Macmillan made rear-driven bicycle		
1863	Velocipede *Boneshaker* bicycle by Pierre and Ernest Michaux	1847: Institution of Mechanical Engineers was founded—to share mechanical engineering knowledge.	1838–1848: Chartism; votes for the working class.
1860 - 1880	A variety of cycles built by various people, these included: monocycle, dicycle, tricycle and quadricycle		
1865	Cycle manufacturing started in Paris—French Cycle Industry.		
1870	The Ordinary Bicycle or *Penny-farthing* of J K Starley and W Hillman is made with tension-wire spokes.	1850: Introduction of golf in England.	1861–1865: American Civil War.
1871	The Ordinary is made with radial-wire spokes of W H J Grout.		
1877	J C Garood introduced tubular frame construction.		
1877	Ball- and roller-type bearings were introduced		
1877	J K Starley made the *Coventry Lever Tricycle*.		
1877	J K Starley made *The Royal Salvo Tricycle* with differential gears.		

Date	Bicycle development	Science, technology and arts	Other socio-political events
1878	Cyclist Touring Club was founded.		
1879	H J Lawson made the *Bicyclette* with rear chain drive.		
1880	Hans Renold made the bush roller chain for J K Starley.		
1881	K White and G Davies made the free-wheel mechanism.	1886: First car was invented.	
1885	J K Starley made the diamond-shaped frame for the *Rover Safety Bicycle*—what we know as basis for the modern bicycle. Eugene Christophe invented toe-clips.		
1886	A J Wilson and D Albone made the *tandem safety bicycle*.		
1887	D Albone made the ladies safety bicycle.		
1888	J B Dunlop made the pneumatic tyre.		1889–1902: Boer War.
1890	Drop handle bars were used for racing.		
1895	National Clarion Cycling Club was founded.		
1897	A M Reynolds and J T Hewitt made butted steel tubes.		
1899	*The Gradient* derailleur speed gear was made.		
1900	Raleigh introduced the all-steel safety bicycle. The cycle motor was introduced.		
1901	The Sturmey-Archer epicyclic hub gear was produced.		1914–1918: World War I.
1903	Marston Sunbeam produced the *De Luxe* safety bicycle.	1922: BBC was formed.	
1911	The Bowden electric lighting generator was produced.		1917: Russian Revolution.
1914–1918	Bicycle production switched to producing modified bicycles for the war effort.	1945: First atom bomb was used.	1929 till late 1930s: The Great Depression.
1925	The concept of the lightweight bicycle. Zefal started producing Christophe toe-clips.		1939–1944: World War II.
1934	Maes designed *Maes Bends* handle bars.	1948: National Health Service was founded.	1947–1991: Cold War.

(Continued)

Date	Bicycle development	Science, technology and arts	Other socio-political events
1935	High-tensile butted tubing became available from Reynolds.	1955: Independent television was formed.	
1938	The concept of the ultra-lightweight bicycle was introduced.		1955–1975: Vietnam War.
1948	Adam Hill—made Hill Special—become the *Father of Lightweight Bicycles*.	1962: Beatle pop group. 1969: Moon landing.	
1963	Cinelli produced aluminium drop handlebars.		
1969	Raleigh Chopper was introduced; 1.5 million were sold. Also made by Schwinn and other manufacturers.	1976: Concorde supersonic aircraft. 1978: Water speed record, 317 mph, by Ken Warmby.	1982: Falklands War
1974	BMX-type bicycles introduced.		
1981	MTB/ATB bicycles introduced.		
1996	MTB became an Olympic sport.	1997: Land speed record, 763 mph, by Andy Green in Thrust SSC.	
2008	BMX became an Olympic sport.		
2012	Cycling became a major attraction at the Summer Olympics in London.	2019: Air speed record, 2,193 mph, in Lockheed Blackbird.	

Chapter 2

Holding it together—the frame

The frame is what holds the bicycle together, over the years bicycle manufacturers have sought to make the frame as light and as strong as possible. The two original designs from the 19th Century are still the basis of most bicycles, that is, the diamond frame as was used on the Rover Safety of 1885, and the cross frame that was used on the 1888 Rudge Bicyclette.

Figure 2.1 Lacquer finished bicycle in Hubei Museum of Modern Art, China.

For clarity, we sometimes use the term frameset. This means frame and front forks. This might include the bottom bracket assembly and/or the headset.

Tech note

Frame materials, construction methods and sizing are discussed in other chapters too.

Function of the frame is to hold the components together. It sounds simple; but it's the juxtaposition of the components, even by moving half a millimetre, that will completely change the bicycle. The basic shape of the frame has changed little since the 19th Century, but it has an effect that changed completely in how it works. It's the same with the production method, simply joining the tubes together, but in many completely different ways. Each and every frame designer and maker think that the product which they make is the best ever, and many of them have riders which will ride no other make. This business is full of passion and skill. There are clubs and gatherings for many frame marques. Of course, to make a frame into a bicycle, the right components must be added, some of the organizations related to veteran and vintage cycles are very passionate about this too.

In current bicycling circles, we have lot of variations, commuting, road racing, touring, time trials, mountain bike, bicycle motocross, track racing, classic and vintage gatherings to name a few. At the centre of this is the bicycle frame.

Figure 2.2 Ribble carbon fibre elite bicycle.

Figure 2.3 Colnago flagship bicycle.

DIAMOND FRAME

This constitutes the down tube, the seat tube and the top tube. Added are the seat stays, chain stays and front fork.

The diamond shape of the three main tubes controls the size and performance of the whole bicycle. The rear stays and front fork allow the connection of the wheels. The diamond shape of the main frame is an important shape in the study of practical mechanics (physics); in the resolution of forces.

The basic diamond frame comes in a variety of guises and variations, for instance, sloping top tube, split seat tube, loop or step-through. However, the mechanics remain the same. The seat tube affects the size, the distance between the saddle and the pedals; in other words, it is a function of leg length. The top tube is about reach, upper body size and arm length. The down tube, which may take the form of a single tube or double tube or loop for step-through models, carries the resultant forces applied to the other two tubes.

This type of mechanical structure can be found on other vehicle in a slightly different form. Norton Motorcycles used it for their famous TT winning featherbed frame and bomber airplanes used it in what is called dihedral construction, so that they could use a non-stressed cloth outer skin.

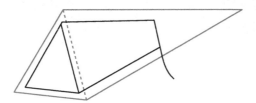

Figure 2.4 Diamond frame.

CROSS FRAME

The cross frame, as its name suggest, is a cruciform shape, simply horizontal and vertical components. The vertical component carries the seat and bottom bracket/pedals in the same way as the seat tube on the diamond frame. It does not have chain stays nor seat stays, the semi-horizontal tube extends past the vertical tube to hold the rear wheel in place. At the front end of this tube is the steering head tube to locate the front forks. This design is not as rigid as the diamond frame, but it has other advantages. It is used by builders of folding bicycles and in a flattened format for some specialised carbon fibre time-trial machines. A version of the cross frame is used in modern light-weight motorcycles; often these designs incorporate the engine as a stressed member. Also, it can be found in a modified form in sports and racing cars, in these the central member is made of a three-dimensional tunnel or tube. Again, carbon fibre is usually the chosen material for extreme lightness.

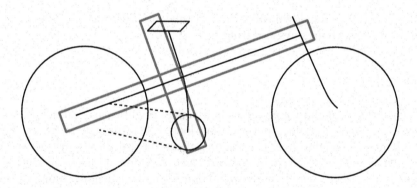

Figure 2.5 Cross frame.

ALTERNATIVE FRAMES

There always has been, and always will be, bicycle builders who want to flaunt the accepted norm for frame design. Their products are usually complex and therefore, expensive.

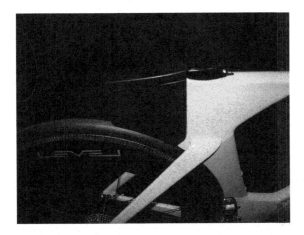

Figure 2.6 Aerodynamic detail.

Dursley–Pederson: Designed by Danish inventor Mikael Pederson and built in the small English country town of Dursley in the West of England. It constitutes a number of semi-flexible tubes and a saddle which is reminiscent of a hammock on flexible mountings. It is soft and comfortable to ride.

Flying gate: Build by Trevor Jarvis of T J Cycles in the industrial Northern City of Bradford, Yorkshire, England. Designed as a time-trial bicycle, it has a very short wheel base and is very stiff.

Loop frame: A variation of the diamond frame, using a large curved tube instead of the top tube. This gives a unisex step-through design. Very popular in Holland for a variety of delivery bicycles.

Folding bicycles: There are a number of folding bicycles; the most popular and expensive are those by Brompton and Dahon. Favoured by

Figure 2.7 Range of folding bicycles.

Figure 2.8 Folding bicycle detail.

city commuters and a variety of other enthusiasts who like to travel to far flung destinations by train and then, cycle.

Electric bicycle frames: The basic electric bicycle frames tend to be based on roadster diamond-shaped frame and made very sturdily. However, there are a growing number of variations including folding models and carbon fibre lightweight styles.

Figure 2.9 Electric bicycles with a variation on a cross frame.

FRAME MATERIALS

Materials and properties of materials are discussed fully in separate chapters.

Steel

There is a lot of different types of steel; but most of the steel cycle frames are made from an alloy using chromium and molybdenum (chro–moly). The tubes may be plain gauge, the same thickness all along or butted, that is, thicker in parts. Aircraft metallurgist Adam Hill discovered that the use of butted chro–moly tubes allowed the frame tubes to be made from much smaller tube. This reduced the tube diameter from 1 ¼ inch to 1 inch (approx. 32–25 mm). So, reducing the weight of the main frame by about 20%. Using light gauge forks and stays gave an overall weight reduction too.

The tables in the material section show the relative weights (density) of the different materials where these are available.

Most elite level steel bicycle frames are made from a version of chro–moly with double- or triple-butted construction.

Steel has a number of advantages:

- It is reasonably readily available, so relatively low cost.
- If not strained, it will last a long time.
- It can be joined in a number of ways, lugged or fillet brazed are the most common—these methods are ideal for low volume production.
- Steel is somewhat elastic and springy; making it responsive to ride and absorb bumps—a feel favoured by many cyclists—clubs exist specially for steel frame riders.
- Steel is the most recycled material, so contributing to maintaining the environment.

Figure 2.10 Specially shaped rear stays for large gear block.

Aluminium

Aluminium is the most common material naturally occurring in the earth, however, it is very expensive to process from its ore, bauxite, to a usable material. Aluminium is usually alloyed with magnesium and copper to give it the properties needed for structural use in bicycle frames. Pure aluminium is too soft for most purposes. Aluminium is very light in weight, for the same equivalent strength as steel; it is about a third lighter. However, aluminium can work-harden, that is break-up after time, especially if subjected to vibrations. This is one of the reasons passenger aircrafts are scrapped after about 25 years. As aircraft production is moving on to carbon fibre composites, there is a surplus of aluminium. Aluminium can be easily formed into bicycle components by a process called hydroforming. The sheet aluminium is put into a mould, and water under great pressure is used to force the sheet to the shape of the mould. Aluminium tubes for bicycle use are not as readily joined as steel. For low volume production, bonding (gluing) is often used; the other method is tungsten inert gas welding. Mass production aluminium frames are machine tungsten inert gas welded. The setting-up of these machine tools is very expensive, so the aim is usually very high volume—not easily achieved with an expensive material.

Riding aluminium alloy frames as the advantage of lightness; but the disadvantage that there is not the same feel as with steel. It depends on your riding style and how valuable weight reduction is. It is readily recyclable.

Figure 2.11 Computer numerical control machined forks.

Carbon fibre

Carbon fibre is now becoming a cheap material as its range of uses has increased. The volume used in the aircraft, super yacht and supercar production means that it is readily available. Production techniques for cycle frames are usually hand lay-up using moulds. When laid–up, the material is usually cured in a autoclave. The autoclave is a sort of oven which has both vacuum and heat. This brings the fibres together and cures the resin. It is possible to do this low volume production, but methods are constantly being improved. Carbon fibre is not currently readily recyclable, so its disposal is an environmental issue. It is very light for its strength, but it is not necessarily strong in every direction.

Lugs

Lugs are usually made from cast steel; joined to the frame tubes by brazing or silver soldering. Custom bicycle manufactures usually shape the lugs to ornate patterns.

Fillet brazed

Fillet brazed are the steel tubes that are brazed without lugs. The bead of brass spelter making a fillet line between the various tubes.

FRAME SIZING AND FRAME FITTING

See Chapter 6 for details on this.

Stress and Strain with an ever-increasing desire for lighter frames, the margins on ultimate strength and loads transmitted by the riders are increasingly being diminished. Some World Tour level teams are now making bicycle frames expressly for the strength of the particular rider.

Figure 2.12 Custom mountain bike with fillet brazed steel tubing–rear suspension.

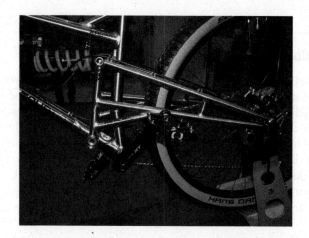

Figure 2.13 Swinging arm mechanism.

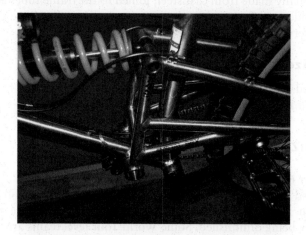

Figure 2.14 Spring and shock absorber mounting.

Figure 2.15 Stiffened seat tube.

Figure 2.16 Front forks.

Figure 2.17 Small chainring and large rear block cassette.

Figure 2.18 Complex bottom bracket and swinging arm set-up.

CLASSIC FRAME BUILDERS

To give the readers some ideas about the number of frame builders, the following list is supplied by Classic Lightweights UK. Each builder considered that each frame they made was the best. If you are interested in making your own frames, it is worth looking at a few examples in this list. If you study the names, you will also recognise that some of the makers have also been/are great riders too.

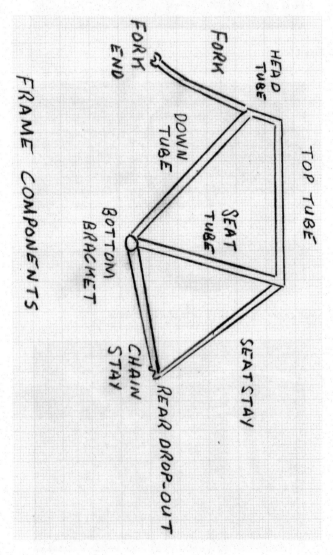

Figure 2.19 Frame components.

A
C W Alexander
Allin
Alpex

B
Jack Baguley
Baines
Johnny Berry
Bianchi: the early years
Bianchi Paris-Roubaix
P Barnard and Son
Bates
C Bertrand
V T (Tom) Braysher
George Brooks
Jim Broome
Buckley Bros
Geoffrey Butler

C
Carpenter Cycles
Carlton Cycles
Geoff Clark
Claud Butler
Claud Butler Bilaminates
Jim Collier
Condor Cycles
Joe Cooke

D
Dave Davey
Harry Daycock
Fred Dean
H M Dickinson
Duckett (A G Duckett & Son)
Duke 20th Century Cycles

E
Vic Edwards
Ellis-Briggs
Elswick-Hopper Convincable
Ephgrave
Ephgrave: A pictorial story
Ephgrave: James Grundy
F W Evans

Excel Cycles Colliers Wood
Excel Cycles Woolwich

F
Flying Scot
F W (Freddie) Folds
Fothergill
Frejus

G
Bill Gameson
Cicli Gasparetto
A S Gillott
Granby
Bill Gray
Walter Greaves
H E 'Doc' Green
Wally Green

H
Evelyn Hamilton
Hamilton-Butler
Pat Hanlon
R O Harrison
Hateley's Lightweight Cycles
Hawkes of Stratford
Jack Hearne
Hetchins
Higgins Cycles
Hill Special–Adam and Denis Hill
Hilton Wrigley
Hinds
Hines of Finchley
Hobbs of Barbican
W F Holdsworth
J A (Jack) Holland
Holmes of Welling
W B (Bill) Hurlow

I
Innanzi Tutto (Bondun's)

J
Jensen Cycles
JRJ (Bob Jackson)

K
Knight Cycles

L
Leach Marathon
F.A. Lipscombe
Norris Lockley
James W. Long
Don Louis

M
Macleans Featherweight Cycles
Major Bros
Major Nichols
Mal Rees
Tom Maysh
Mercian Cycles
Merlin–the early years
Merlin
A F Mills (Welling)
Moorson Cycles
H R Morris
H R Morris lugs
Sid Mottram

N
Nelson Cycle Company
Nervex Professional Lugs

O
Walt Ormsby

P
Paris/Rensch Cycles
Paris catalogue 1948
Theo Parsons
Pemberton Arrow
Pennine
W W (Bill) Philbrook
Len Phipps
W & E Pollard
Cliff Pratt
Frederick Pratt
Stuart Purves

Q
La Querée
T J Quick

R
Raleigh (track frames)
Raleigh SBDU
Les Rigden
Rivetts
Rondinella
Rory O'Brien
Rory O'Brien (2)
Rotrax
Dave Russell

S
Sanders, F J
Saxby's
Saxon
Selbach
Shorter, Alan
Shrubb, Cliff
Jack Sibbit Cycles
Jim Soens
John Spooner Cycles
Stevenson, Charles
Clive Stuart Cycles
by Curt Yamamoto (Seattle)
Success
P T (Percy) Stallard

T
Jack Taylor
Thanet

U
The Upton–J J Cooper

W
Walklings
Waller
Welded/lugless frames
Whitaker & Mapplebeck (Pennine)
Williamson Brothers
Wilsons of Birmingham

Wilson Cycles, Sheffield
Witcomb
Woodrup
Wren, Cyril (and Clubmans Cycles)

Y
Youngs of Lewisham

SIMPLE GUIDE TO FRAME DESIGN

1. Choose seat tube length, usually based on leg seam size less than about 10 inches (25 cm).
2. Choose top tube length, usually the same as the seat tube and depending on your body size.
3. Choose seat tube angle, between about 71° for a laid-back style and 74° for a race model.
4. Choose the bottom bracket height, this affects the centre of gravity. For a track racing frame, it will be higher to allow lean on the banking, while for a long distance tourer, it will lower to give easy stand over height.
5. Choose the wheels that you are going to use, find the radius of them.
6. The front fork length and rear stay length can now be drawn in.
7. Finally, you can add in clearance for mudguards and brake fixings.

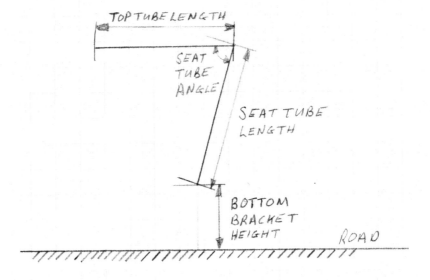

Figure 2.20 Frame design 1.

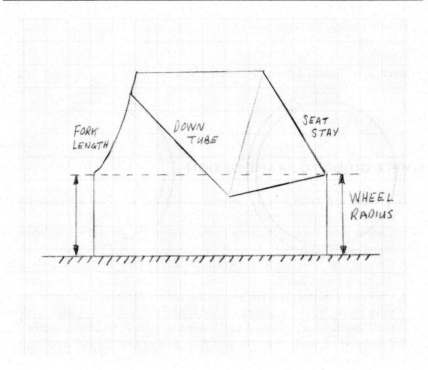

FORK
LENGTH

DOWN
TUBE

SEAT
STAY

WHEEL
RADIUS

Figure 2.21 **Frame design 2.**

Chapter 3

Road holding—wheels and tyres

REQUIREMENTS

Bicycle wheels have the job of enabling the machine to roll along the road. The front wheel carries out the steering function; the rear wheel transmits the drive to the road. This has been the situation since the invention of the bicycle. The sizes of the bicycle wheel have not changed much since the invention of the Draisienne. The original Draisienne had 27-inch diameter wheels; the design of the Rover Safety reduced the wheel size to 26 inch.

It can be imagined that the size of the Draisienne wheel was chosen as the size that the designer could comfortably stand over. It could also be the size which was used for carts and jigs used in agriculture in Germany. The reduction of the diameter for the Rover Safety can be taken from its name—smaller probably equals safer. It was also designed for use by women; women are statistically shorter than men. Also, English men from the Midlands were shorter than many European races. Noticeably, Dutch people and bicycles tend to be at the top end of the height distribution curve.

There are also 29-inch wheels which have found favour in MTB usage, and smaller diameter wheels for children's bicycles. The 29-inch wheel gives advantages when rolling over uneven ground.

Tyre width and aspect ratio: The width of the tyre is very important. Noticeably, the width of the tyres on sports and racing bicycles has increased marginally over the recent years, this may be simply a fad, alongside disc brakes or it may be because a lot of professional road race events have started using long stretches of unmetalled, for example, the Paris-Roubaix and the Strada Bianca. The Paris-Roubaix is 257 kilometres long with 54.5 kilometres of pave or cobbles. The Strada Bianca is 184 kilometres with 60 kilometres of dusty white gravel.

The height of bicycle tyres is usually approximately the same as the tyre width. This ratio is known as the aspect ratio.

Aspect Ratio = height of tyre / width of tyre

You will see some small variations in this, but not like on car tyres where the aspect ratio can vary between about 0.35 and 1.0.

Tyres on bicycle wheels are fairly forgiving in fitting sizes, but please be aware that there are four completely different fitting standards in use.

ISO: The International Standard; this uses a two-digit number, a hyphen and a three-digit number, for example, 28-622. This tells you that it is 28 mm wide and 622 mm in diameter.

Imperial: For example, 26 × 1⅜ this is 26 inch diameter and 1⅜ inch width.

Metric: For example, 700 × 25 C, which means 700 mm diameter by 25 mm width.

American: For example 26 × 1⅜ which is similar to the Imperial size, but sometime you may see the width written as a decimal, 26 × 1.375.

Tech note

It should be noted that although some organizations talk about equivalent sizes, these are only approximate at the best and that the retaining method should also be compared.

NUMBERS

Bicycle tyres are very much a consumer item. They are essential; but when we are fitting an expensive pair to a bicycle to be used for an event, it is worth considering the numbers of them in use at any one point. For example, in China there are about 600,000,000 bicycles, that means 1,200,000,000 tyres. They are likely to fit more than one new pair each year, which means making more than three million tyres a day just to maintain current needs.

RADIUS OF GYRATION AND MOMENTUM

Newton's first law tells us that a body will remain at rest, or continue to move in a straight line if no resultant force is acting upon it. Applying this to a bicycle wheel, it will remain stationary unless we move it and it will keep rolling unless it is stopped by something. If you put the bicycle in the workshop stand, the wheels will turn a little until the valves are at the bottom. Now spin the wheels with equal force, the front wheel is likely to spin for longer than the rear wheel. The resistance of the front wheel is generally only bearing friction; but with the rear wheel, there is extra friction caused by the free-wheel cassette.

Thinking about riding the bicycle, we have four different situations with regards to the wheels, these are:

- Accelerating away from rest
- Maintaining a constant speed whilst riding
- Cornering
- Stopping

These four situations have a relationship to Newton's first law. They are all about momentum.

Linear momentum of the bicycle is expressed as mass (m in kg) multiplied by the velocity squared (v^2 in m/s):

Momentum $= m \times v^2$

As 1 kg m/s^2 equals to 1 Newton, the answer can be expressed in units of force. So, we can calculate the force on the pedals to overcome the momentum for both acceleration and maintaining a constant speed—especially interesting for hill climbing.

Moment of inertia (I): It is often used in solving problems involving rotating bodies, for examples, wheels. Moment of inertia is also called moments of area and moments of mass. For example, if a mass is δ m and the wheel rim is rotating on a radius of r, then:

$r\delta m$ = first moment of inertia

$r^2\delta m$ = second moment of inertia

If the wheel is made up of several masses, such as with disc brakes, then the individual parts need to be summed up such that:

$I = \Sigma r^2 \delta m$

Another way of dealing with the moment of inertia (I) of a wheel, viewing it as a plane disc, is to consider it in terms of an RMS value, in other words, using the constant (k) of 0.7071. The constant 0.7071 coming from the sinusoidal wave form. So, sum the total mass (M) and multiply it by K, which is 0.707 times the outer radius R.

$I = MK^2$

where, I is moment of inertia, M is mass and K is outer radius times 0.7071. So, the inertia is increasing directly proportional to the mass and as the square of the value of K. As this is a comparative calculation to compare wheels, not absolute values are needed, so it is the easiest solution.

Weighted wheels: There has been a school of thought in time-trial racing to use weighted disc rear wheels. The weights were available with various values of mass. The concept being that of maintaining momentum like a flywheel. Given a flat out and home course with the absence of cross wind, this helps the rider to maintain a continuous speed. They flatten out the minor irregularities that any rider gets with a normal chainset where one pedal is at top dead centre, and the other at bottom dead centre. In cross winds and circuits where there are any significant climbs, this advantage may not be significant.

Fixed gear wheels: These give similar advantages to the weighted wheel in maintaining constant speed. The wheel and chainset do not have any freewheel position, so that they must continuously turn together. This means that the rider cadence rate—pedalling speed — must change with variations in road speed. The old-school time-trial rider would practice on both the rollers and the road to maintain a particular cadence to achieve a particular time for a distance using a chosen gear. All clubs used to have low-gear time trials at the start of the season to encourage high-speed pedalling. This was fixed gear of 72 inch maximum, riders used between 65 inch and 72 inch. For normal events in the season, a fixed gear of 82 inch or 85 inch was typical.

RIM AND TYRE TYPES

There are four different types of rim and tyre fittings and within these four, there are a number of variations.

- Wired-on; high pressure, Endrick and Westwood
- Clinchers
- Tubeless; a version of clinchers
- Sprint/tubulars

Wired-on: These are built in the same way as car and motorcycle tyres, in that they have a wire bead at the bottom of each of the tyres' two walls. The wire of the bead is coated in rubber and in effect forms the end of the wall. The diameter of the wire bead is same as the inner part of the rim, so that the bead seats on the rim to hold the tyre in place. The sides of the rim prevent the tyre from coming off. To fit the tyre, the bead has to be stretched over the sides of the rim, this can be quite a challenge and skill is needed to fit them without damaging the tyre or the inner tube.

Clinchers: These tyres are also referred to as folding tyres. With the wired-on tyres, the wire bead must not be bent or the tyre will be damaged, so taking a spare tyre on a long journey is not practicable; hence the development of the folding tyre. The folding tyre has a bead, but the bead is flexible and can be folded without damage. Instead of the bead sitting on the base of the rim, like the wired-on; the folding bead catches on a hooked edge on the inside of the rim. These tyres require much less effort or force to fit them, but they still require skill or technique.

Inner tubes: Both wired-on ones and clinchers use inner tubes. This is a tube to contain the air to inflate the tyre. There are three types of valves in use:

- Woods valves: These are rarely seen now. They have a removable valve core and maintain pressure by having a self-sealing design. These are used on utility type bicycles which run at low pressure.
- Schrader valves: A screw in core as used on most cars and motor-cycles. These are used on all MTB and some road bicycles.
- Presta or high-pressure valves.

Tubeless: The current move towards tubeless tyres is due to the following advantages they offer:

- Less weight: Less material, so less mass.
- Easy to replace the tyre: No inner tube, so a very simple job and no need to worry about damaging the inner tube.
- Can be used with folding tyres.
- When filled with solution, they become self-repairing. Following a puncture, the tyre is re-inflated if pressure is lost, this may be done in seconds using a CO_2 canister.

Sprint/tubulars: Often called sprints and tubs. The rim hasa slightly concaved outer surface on which the tubular sits on. The tubular is exactly as its name describes. The outer tyre walls are stitched together to form a tubular shape. The inner tube is inside this tube. Should you have a puncture, then you must undo the stitching to remove the inner tube and then re-stitch it again while being careful not to prick holes in the tube with your needle. With care it is possible to run the punctured tyre, when removed from the rim and inflated, in a basin of water to see the bubbles coming from the place of the puncture. Then only undo a small section of the sewing to enable the punctured part of the tube to be removed and repaired with a patch before re-sewing.

INFLATORS

Inflating tyres have always been an issue; there is no quick fix if you haven't got access to an inflator of one type or another. The four inflators are:

- Track pumps
- Portable/frame fit pumps

- Gas canisters
- Compressed air from compressor

The track pump is probably the best system, good and heavy duty with a built-in pressure gauge. Remember to put it in your car boot if you are driving your bicycle to an event or outing. The portable or frame fit pumps are ok, but may fall off the bicycle on bumpy roads. It is a good idea to secure them with tape. The CO_2 gas canisters are much easier to carry than a pump and will re-inflate the tyre in seconds—but always take two as they are one-shot use only.

TYRE PRESSURES

Most tyre manufactures state a maximum pressure; this should not be exceeded as the extra pressure may damage the tyre by stretching. Finding an ideal tyre pressure for an event or particular circuit is to some extent, a personal preference related to bicycle handling style.

A LESSON FROM CAR RALLYING

When I was preparing a rally car for a major event, the very famous driver, who'll remain nameless, wanted to get his tyre pressures to suit his driving style. So, we hired a circuit and started at 8:00 am with one set of pressures. Going through the day, the pressures were changed, and lap-times and handling performance were noted. The notebook was filled. At 5:30 pm, the driver came into the pits for the last time—he said, "they are perfect now, so I do a final check". Yes, they were the same as they were at 8:00 am.

ROLLING RESISTANCE

Rolling resistance is the energy that is lost when the tyre is rolling along the road. This is accounted for by the constant deformation and straightening out of the tyre walls and other parts. The factors that affect rolling resistance are:

- Tyre pressure
- Tyre diameter
- Tyre width
- Tyre construction
- Tyre tread

Larger diameter tyres, for instance 29ers, roll more freely as the curve is less pronounced. Small diameter tyres, as on folding commuter bicycles, have greater resistance. Tyre construction is very important as the amount of actual material to be deformed will increase the rolling resistance. Hence, the use of sprints and tubs especially for time trials, these are made very low mass. Wider tyres, at the same pressure as narrower ones, roll with less resistance than the narrow ones. This is a balancing compromise for racing or other high speed events off-road—such as Audax and Sportives. For maximum speed, the very narrow racing tyre pumped to maximum pressure will give the least rolling resistance, but the constant jogging and jolting will both exhaust the rider and possibly damage the bicycle. It will also be at the greatest risk of a puncture. A wider tyre of maybe 28 mm with a slightly reduced pressure will roll more freely, give a more pleasant ride and be less likely to puncture.

Table 3.1 Typical tyre rolling resistance data

No.	Width	Weight gram	Rolling resistance at 120 psi watt	Rolling resistance at 60 psi watt	Type
1	25 mm	240	7	9	Racing–tubes
2	25 mm	255	11	15	Racing–tubeless
3	25 mm	290	13	18	Racing–tubular
4	35 mm	335	n/a	24	E-bike–tubed
5	2.25 inch	645	n/a	20	MTB/ATB–tubed

Racing road bicycle tyres tend to run at about 100/120 psi, MTB and E-bike tyres 55/65 psi. I know that if I have a puncture, especially in the rear wheel, I can feel the resistance increasing as the tyre goes flat. As the tyre deflates, the weight is transferred to the rear wheel as the balance shifts. On very steep hills, the same effect can be noticed; I have occasionally got off to check the rear tyre.

Tech note

At speeds above even 20 mph/32 kph, the rolling resistance of the tyres is negligible compared to the aerodynamic drag.

NOISE

Tyre noise is an interesting subject for study, there are a number of master's/doctor's degree papers written on the subject. Noise is generated by:

* Tyre tread deformation
* Road surface

- Tyre rolling distortion
- Vibration in the bicycle forks and frame

I must admit, one reason that I like riding on sprints and tubs is the high-pitch noise that these generate on a smooth metalled road. It is like angels singing to my ears.

PUNCTURE RESISTANCE AND PUNCTURE FACTOR

Punctures are a bain in the life of cyclists, especially in wet weather when the rain water seems to lubricate bits of glass and chips of flint to get easily into the tyre. My particular problem in Kent is the large number of pot-holes, the impact is like a hammer blow to the tyre which forces in the jagged edge of the road surface to cause a puncture. The standard test for puncture resistance is to hold a 1 mm diameter steel rod against the tyre tread and find the weight (mass) needed to cause a puncture. This is best done using a jig set-up, convert the result to newtons, that is, mass times G (9.81) this is referred to as puncture resistance. The puncture factor is the product of the puncture resistance (N) and the tyre thickness in millimetres. As this is purely empirical and comparative, you may use any units of your choice.

To try to beat the puncture problem, tyre manufactures use a variety of materials both in the tyre construction and as a separate layer. Kevlar is very popular. Of course, it is a good idea to check the tyre visually before or after each ride, removing any flints or other materials which become embedded in the tyre.

TREADS

There is no legal requirement for bicycle tyre to have treads of any kind. MTB tyres have treads to give obvious grip. On road bicycle tyres, the cross-hatch tread served to indicate the amount of rubber which is worn off, so giving a guide to the need to replace them.

FRICTION

The coefficient of friction μ between the tyres and the road is obviously of great importance.

μ (the Greek letter Mu) = Force / Weight

That is the force needed to slide the tyre over the road surface as a proportion of the weight—that is the force normal to the road surface.

Tech note

Weight for most bicycling calculations means mass x gravity in kilogrammes (kg) or pounds (lb). A kilogramme force approximately equals 10 Newtons (N). 1 kilogramme equals 2.205 lb.

The coefficient of friction between bicycle wheels and the road is, I would describe as tentative, for much of the time. That is, the contact patch is so small and therefore, the variations of the road surfaces, even on velodromes, is vast. On a racing bicycle, the contact patch is nothing more than a thick pencil line. The width of the contact patch does not affect the coefficient of friction. The rubber compound and the road surface composition are what effects the figure.

Typically, a sports bicycle tyre on a smooth dry concrete surface will give a coefficient of friction of between about 0.7 and 0.9. On a wet surface however, the figure could be as low as 0.1. The problem is not about skidding when accelerating, or braking, as it is with a car, but about remaining upright. Should the tyre not be able to prevent sideways motion, the bicycle will fall sideways. The area where the coefficient of friction is probably most important is track racing in a velodrome, where there is steep banking offering great opportunity to slide downwards. A possible scenario is where a sprinter rides to the top of the banking and stops, in a track stand, playing cat and mouse with the other riders, before using the slope of the banking to help accelerate across the line. This requires a high coefficient of friction on the shoulders of the tyres as well as the centre section.

Any coefficient of friction over unity (1) means that the tyre is like chewing gum sticking to the road surface. This is exactly the case with some racing car and racing motorcycle tyres. The tyres are heated up so that their treads are sticky, this way the coefficient of friction is raised to about 1.1 or even 1.15. The high levels of road contact are needed for two particular reasons—stopping a 1,500 kg car sliding sideways with lateral acceleration of about 1 G, and to transmit 1,000 HP to the road without spinning the wheels. Before a race starts, the tyres are put into a sort of electric blanket to warm them up, on the opening laps you'll see both racing cars and motorcycles zig-zagging across the track, and this helps in keeping the tyres warm.

With roadster bicycle tyres and those on MTBs.

RIMS

The lighter the rim, the less the moment of inertia, so less effort for the same speed. Originally rims were made from plain steel—low carbon steel or malleable iron—which was easy to shape. The earliest bicycles with artillery wheels—wooden rims and spokes—were fitted with iron or steel tyres. The iron used was the same as used by the farrier for making horse shoes. Then the move to plain steel followed by a variety of other materials, particularly aluminium alloy and most recently carbon fibre.

Rims can be made in several different ways, the most common being extruded from billet, chopped and butt welded. If you look at rims closely, you will see the butt joint—done electrically. Often the makers decal (name transfer) is put over the joint to cover it from the eye.

Besides providing a location for the tyres, the rims may provide a braking surface for the brake blocks. With track bicycles and bicycles with disc brakes, this surface is not needed and therefore, providing the opportunity for more aerodynamic shapes of rims.

Tech note

To reduce weight and improve wearing properties, both horse shoes and bicycle rims have followed each other through the ages in the types of materials used. Both current horse shoes and bicycle rims are available in carbon fibre composite materials. They are both light in weight and can offer a comfortable ride.

SPOKES

There are three main spoking patterns: radial, tangential and semi-tangential. The names give the descriptions of the patterns. The semi-tangential patterns may be one-cross, two-cross or rarely three-cross. That is the number of other spokes, each spoke crosses another.

Spokes are made from high tensile strength wire; this may be round or a more aerodynamic oval shape. Some spokes are butted at each end. Several different finishes are available such as zinc galvanised, chrome, stainless steel or black coated.

The spokes work together as a form of disc, in both tension and compression. The rim needs to be held by the spokes centrally on the hub. When large derailleur free-wheel cassettes are used, like 9 speed; this means that the rim is dished by a large amount, so there is a fragility in the wheel. The number of spokes used on each wheel varies between 18 and 40 for most applications.

HUBS

The hub attaches the wheel to the frame and forks. There are several different bearing arrangements, the two main are the use of cones and separate steel balls or small pre-assembled units. The latter my use ceramic balls for extreme manufacturing accuracy, giving almost zero friction. In any case, bearing friction is in a very minimal amount.

Hubs may be attached to the frame and forks using:

- Plain nuts
- Wing-nuts
- Over-centre quick-release—enclosed cam, also known as Campagnolo type
- Turn and lock—exposed cam

The width of the hubs varies considerably so that the frame and forks need to be able to accommodate them. This is especially so with the use of disc brakes.

Figure 3.1 Foldable tubeless tyre.

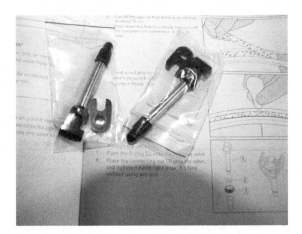

Figure 3.2 Valves for converting to tubeless.

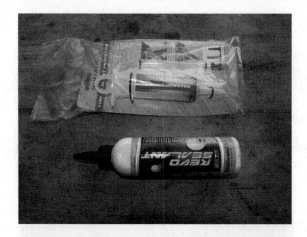

Figure 3.3 Tubeless tyre sealant and injector.

Figure 3.4 Deflating old tyre.

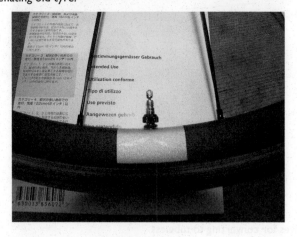

Figure 3.5 New valve inserted with rubber seal.

Chapter 4

Making it go—gears and drive

Gearing is about how far will one revolution of the pedals make the bicycle go. Typically, this is 216 inch to 86 cm. These figures on their own do not mean much to most cyclists. This is the distance covered by a 69-inch gear, or 46 × 18 on average wheels. The 69 inch comes from the size of a typical Penny-farthing front wheel. Penny-farthing front wheels are directly driven by the pedals, so the wheel size altered the gearing, they were made between about 56 inch and 69 inch diameter. This is actually a comfortable gear ratio, and the ratio used on most single-speed bicycles is in this range. It is usually about the middle of the gear ratios used on multi-speed bicycles. The gear table in the appendix gives a summary of gears available based on common chainwheel and sprocket combinations with popular wheel sizes.

The gear size is worked out by dividing the number of teeth on the chain ring by the number of teeth on the sprocket and multiplying the answer by the wheel diameter. For example, with a 46-tooth chainring, 18-tooth rear sprocket and 27 inch wheels, it is:

$$46 / 18 \times 27 = 69 \text{ inch}$$

Be aware that this is not a ratio, it is a comparative gear size. A touring bicycle may be equipped with 20 gears between about 31 inch and 108 inch, an elite road racing bicycle may have a range of 22 gears between 64 inch and 132 inch.

MECHANICS OF GEARS

Mechanical advantage: One of the purposes of gears is to give a mechanical advantage, that is, a 10 N force can move a 100 N load with a 10:1 gear ratio. This will mean a movement ratio of 10:1. Moving the pedal 10 cm will only move the load 1 cm. But with this, a 10 N effort can move a 100 N load. This way, with suitable gearing, you can ride a fully-loaded touring bicycle up the steepest mountain road. A very small chainring with a very large rear sprocket and your feet turning quickly.

Speed ratio: The opposite of the mechanical advantage is the speed ratio. This originally started to be used in time-trialing. Chainrings were made bigger and bigger, driving the smallest rear sprocket, so that the road speed could be increased for any pedalling rate on flat roads, or with a tail wind.

GEAR RATIOS

This is the ratio of the number of times, the rear wheel turns in comparison to the chainring. With our 46/18 set up, this is:

46 / 18 = 2.555

To get the same ratio with a larger chainring, we would need to increase the sprocket size too. If we reduced the chainring size, we could reduce the sprocket size too. If you study the gear tables, you will be able to see the relationship.

Figure 4.1 **Gear cable inside chain stay.**

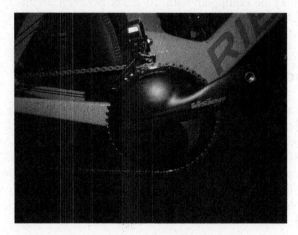

Figure 4.2 **Dome shaped chain stay to give good aerodynamics.**

Figure 4.3 Close up of front end.

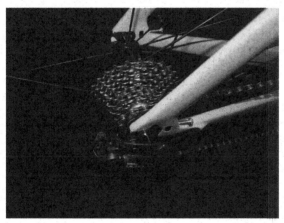

Figure 4.4 Mechanical wide range rear changer.

Figure 4.5 Rubber toothed belt drive.

CHAINSET

The chainset comprises cranks and chain wheels. This attaches to the bottom bracket spindle to provide a pivot point and the chain wheel rings—the parts with the teeth are called the chain rings. The rings transmit the power to the chain.

Chainsets come in a number of different types, fittings and materials; as well as chainring sizes and crank lengths.

BMX bicycles tend to use steel one-piece set-ups. The crank and chainring and bottom bracket are made as an assembly.

Elite bicycles use aluminium alloy chainsets with interchangeable chain rings. Other sports and commuter bicycles tend to use cheaper steel chainsets; these may have interchangeable rings.

Chainrings vary in fitting patterns—either three or five bolt configurations. The number of teeth varies between about 28 and 54. The gear tables in the appendix show you how these alter the gear sizes.

The cranks are attached to the bottom bracket, usually in one of three popular ways described below:

- Round with cotter pins
- Square taper
- Splined drive

To remove the cotter pin type, the nut is loosened and the pin knocked out. You need to be careful not to damage the thread on the cotter pin. One method is to loosen the nut and take it to the end of the thread, then use a soft metal—brass or aluminium—and drift between the pin and the hammer.

To remove the square taper and splined drive, types first remove the retaining nut, and then use the correct removal tool to pull the crank off the bottom bracket spindle.

BOTTOM BRACKETS

The bottom bracket is attached to the frame in one of two ways:

- **Screw fit:** Also known as Birmingham small arms screw type or as a British bottom bracket. These are used on about 90% of all frames, no matter where they are made. The cups screw into the threads on the frameset. Be aware that one side has a right-hand thread, the other a left-hand thread.
- **Press fit:** As the name suggests these are pressed into the frame. This type are usually found on carbon fibre frames, where the bottom bracket shell has an insert bonded to the frame tubes. To fit these, it is best to use a hydraulic press, although the job can be completed with a hammer and suitable mandrill.

Tandems may have bottom brackets which are eccentric in the cups, this allows the front and rear chainsets to be adjusted in position relative to each other; this enables the chain tension to be set.

CRANK LENGTH

The crank length is measured from the centre of the bottom bracket spindle to the centre of the pedal axle. The common sizes are: 160 mm, 170 mm and 175 mm. The longer the crank in effect, the greater the leverage that you can apply; same as using a tyre lever or crowbar. The longer it is, the more the turning effort—torque—you can apply. Longer cranks are pre-ferred for hill climbing as these, in effect, lower the gear ratio—they also have the psychological effect of making the rider feel stronger.

Longer cranks are also more comfortable for riders with longer feet—they allow a more relaxed muscle movement of the ankle; whereas riders with shorter feet may have a faster pedalling rate. There are also reverences in the various cycle training literatures to a ratio between leg length and crank length. However, people with longer feet tend to have longer legs, so the rule of longer feet use longer cranks applies. More importantly, to both the casual and serious cyclist alike is the pedalling style.

Pedalling style is something which is often overlooked by riders. If the ball of the foot, that is, where the toes pivot, is central over the pedal spin-dle, the angle can be used to both increase the cadence—pedalling rate—and give more effort onto the crank.

The force in Newtons (N) which the rider applies to the pedal multiplied by the crank length in metre (m) gives the torque (Nm) which is being devel-oped. For example:

1000N force on pedal × 0.17m crank length = 170Nm torque

Figure 4.6 Super small single chain ring.

Figure 4.7 Exceptionally wide chain angle.

BLOCKS OR CASSETTES

Gear blocks or cassettes, come in a variety of fitting systems and tooth ranges. The original types were referred to as blocks as they were a set of sprockets screwed together, which in turn screwed onto the rear hub. The current cassette systems fit on a spline on the rear hub, taking their name from the now redundant music and video cassettes, which used a splined drive mechanism. The popular systems are made by Shimano and Campagnolo.

Figure 4.8 Very long arm with wide range cassette.

Figure 4.9 Extreme wide range gear set-up.

Figure 4.10 Wide chain angle 2nd view.

CABLE SYSTEMS

There are two main cable systems, these are:

- **Friction levers:** Such levers have a thumb screw which can be tightened-up to hold the cable in the required gear position. Good quality systems work very well. Sometimes they are enhanced with a rachet type friction surface. The rider needs to feel the gears engage and the friction keeps the selected gear in place.
- **Index levers:** Such levers have stops in the housing, so that it is just a matter of flicking the lever to select the next gear.

The current elite type of index systems, usually operated by moving the brake lever sideways as a double movement. That is each gear can be selected in turn, or with a rapid movement a wider gear change—say from top to bottom—can be achieved.

Figure 4.11 Handle bar end gear lever.

ELECTRONIC SYSTEMS

Basically, electronic systems are just like cable systems using servo motors to move the mechanical parts. However, the real technology is in how a rider can set them and use the data. For example:

- Given pre-selected gear ratios; movement of the front changer, say changing up, the control will lower the rear sprocket at the same time so that there is not a stepped gear change.
- If set up with a GPS system, such as Strada, you can analyse the chosen gear and speed at any point in your ride.

Figure 4.12 Electronic rear changer with wire in stay.

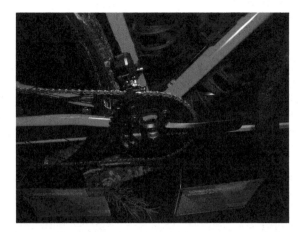

Figure 4.13 Electronic front changer.

Figure 4.14 Very clean-cut electronic gear set-up.

WATT METER

This records the rider power output in watts at any point of time. Watt meters are available to be fitted to, or part of:

• The pedal construction
• The crank construction
• The bottom bracket assembly

PEDAL AND CLEAT SYSTEMS

There are a number of pedal and cleat systems as well as a range of toe-clips and straps. The older toe-clip systems also could be used with shoe-plates. The shoe-plates were nailed to the soles of the cycling shoes, so when the strap was then tightened, the shoe could not be pulled out of the pedal. This enables the foot to be able to lift the pedal and pull backwards as well as pushing down forwards. The problem with toe-clips is that the rider has to reach down to release the strap before being able to withdraw the foot when coming to a halt. This could be a tricky manoeuvre in traffic, it often leads to riders doing track-stands to avoid falling on the road or worse, still onto adjacent vehicles.

The cleat system—also known as clipless—uses a spring-loaded mechanism which locks the shoe in place both vertically and fore/aft movement. Twisting the foot a few degrees, allows the shoe to be removed without the use of the rider's hand. When coming to a road junction or other point where you might need to put a foot on the floor, the technique is to twist your foot to free it from the pedal, and then return it to the pedal at a slightly twisted angle so that it will not engage the cleat and pedal. In that way, if you need to put a foot on the road it is quickly done; or if you are able to proceed, just put your foot straight and as you press down, it will automatically engage. When you use a cleat system for the first time, you might find it unnerving. Before you use a new cleat system on the road, adjust the tension of the spring mechanism with the Allen, so that it is both secure and easy to turn without straining your ankle.

The two most popular systems are Shimano SPD and the Time system. Shimano make a double-sided SPD pedal which is great for MTB use as the rider doesn't need to check the pedal position—this pedal is also great for winter road use.

The pedal systems require suitable shoes. The fixings for the cleats are built into the shoes with an adjustable screw fixing system. So, it is advisable to ensure that shoes and pedals match.

FIXED GEAR

This system is popular with bicycle couriers and some club cyclists. The system is, there is no free-wheel mechanism, so that the wheel and pedals turn all the time. It has a number of advantages:

- Very simple and therefore, reliable.
- Light weight.
- Allows both acceleration and deceleration to be controlled by the legs.
- Good for training as it does not allow the legs to rest.
- Encourages the use of different cadence—leg spinning/pedalling—rates, especially useful for sprint training.
- Gives better control on icy/slippery roads, rather than applying brakes.
- Especially popular for winter training, less components to clean after a wet ride.

CHAINS

The bicycle chain has changed in detail since its first invention. The rollers sit between the links, the two important measurements are the **pitch** and the **width**. The current most popular pitch, that is, the distance between the centres of the rollers is half-inch (1/2 inch). For track use, chains of one-inch (1inch) pitch are sometimes used because they are stronger to withstand the accelerating forces applied by track racers. One-inch pitch is sometimes used by time-trial riders that use fixed gear.

When width is measured, there are two dimensions; the inside width for the sprocket to fit in and the maximum outside width of the rivet. Sprockets come in two standard widths; this also applies to chainring tooth width. The two dimensions are either 1/8th inch or 3/16th inch. The 1/8th inch is used on single speed bicycles including BMX style. In fact, the volume of use of 1/8th inch chains and sprockets on BMX has ensured that this system is available for the very small number of fixed gear riders—which includes city couriers who like to ride fixed gear for the greater control that this gives in traffic. The riding style used by couriers is to use legs only, for both going and stopping.

The 3/16th system came into use with multi-gear bicycles, what was originally know as 5-speed and 10-speed bicycles. The chain and sprocket are narrower to reduce the space needed for the extra gears.

The move to increase the number of rear sprockets has led to making the chains narrower to reduce the space between each sprocket. It has not been possible to make the sprockets narrower as this would make them too weak. However, a fraction of a millimetre has been shaved off the outside of the chain.

Table 4.1 Chain outside width for number of sprockets

Number of rear sprockets	Nominal chain width in mm
6, 7 or 8	7
9	6.5
10	6
11	5.5
12	5.3

Chains are made from high tensile strength steel, which is heat treated and surface coated to prevent corrosion. The narrower chains used on elite level bicycles are made to very close tolerances with high-quality finish and are therefore, much more expensive.

It is possible to use a high-end 5.3 mm chain on a 6-, 7- or 8-speed sprockets; but because it is narrower, it will have a tendency to slip down between the sprockets and jamb. So, it is both a hazard and a waste of money.

MAGIC LINK

These are also known as split links. The link has a spring component to retain the outer section, so that the link can be both assembled and removed without the use of a special chain tool. It is a good idea to keep a spare magic link to the tool-roll.

Rolling radius

When the bicycle is loaded the tyre walls will deform, thus reducing the outer diameter of the wheel and tyre. This distance from the centre of the axle to the road surface is, therefore, less than the nominal radius of the wheel. This will affect a small change to the calculated gear ratio and also the calculations on some bicycle computers. It must be factored in to any time-trial calculations which are done using by this method—particularly long events such as 12 and 24 s as a very small percentage error could ruin a record attempt ride.

TRACTIVE EFFORT

Tractive effort—TE is the force in newtons (N) between the driving wheel and the road. It is another reason for the choice of 27 inch/700 c wheels. Given any torque figure, the TE will be higher the smaller the diameter of wheel and hence, the easier it will be to lose traction.

CALCULATIONS

Originally, the only way to work out your power and pace was to do this manually using time, distance and changes in elevation.

Using an Ordinance Survey Map, you can calculate changes in elevation. Given the weight (mass) of bicycle and rider and the change in elevation using the formula force times distance, you have work done, divide this by time taken and you have power in watts. Of course, you can now read this off on your watt meter on the handle bars.

Figure 4.15 Hub gear set-up on steel framed bicycle.

Chapter 5

Stopping—brakes and braking

We have a wide range of braking systems available for bicycles, the very first bicycles did not have brakes. To stop a Draisienne, the riders dragged their feet on the floor; on a Penny-farthing, riders back pedalled, as you would do with a modern fixed gear bicycle. The Penny-farthing riders started to take their feet off the pedals to go faster downhill, unable to catch their feet back onto the pedals to slow down, a few crashes and then the spoon brake was introduced. This is literally a spoon-shaped piece of metal that presses against the tyre tread—remembering that tyres were solid.

According to the UK law, and it's universal in most of the world, a bicycle used on the road to should have two independent braking systems. This means, on a standard bicycle, one break on the front wheel and other on the back wheel. For the back wheel, a fixed wheel (fixie) allows pedalling resistance to suffice instead of a normal braking system.

Before the introduction of Bowden Cables, rod braking was used; this was used on roadster bicycles up until the 1980s. These were used with Westwood rims, the brake blocks pulled against the inside of the rim. The Westwood rims were designed by Fredrick Westwood in 1891, a Birmingham Engineer.

The Bowden cable, which is widely used for bicycle brakes and gears, was invented in about 1896s by Ernest Mannington Bowden. The concept being that the inner cable pulled against the outer to operate a mechanism. The early Bowden cables were inner metal strands inside an outer wound metal. The problem was that they easily rust and seize. Current Bowden cable uses inner and outer layers of plastics or Nylon; this reduces friction and prevents corrosion.

MOMENTUM

Momentum is the product of mass and velocity. It is a measure of the amount of energy possessed by a moving body

Momentum = mass × velocity

Kg.m/s

So, given a bicycle and rider with a mass of 85 kg riding at 20 kph (12 mph). 20 kph (5.6 m/s)

$$= 85 \text{ kg} \times 5.6 \text{ m/s}$$

$$= 476 \text{ kg.m/s}$$

Tech note

By comparison a large truck of 40,000 kg at 20 kph

$$= 40,000 \text{ kg} \times 5.6 \text{ m/s}$$

$$= 224,000 \text{ kg} \cdot \text{m/s}.$$

To stop the bicycle and rider, brakes must overcome the momentum, which leads to a **stopping distance**. If the bicycle was to stop dead, then the rider would be over the handle bars. The stopping distance is related to the braking efficiency, which is related to the co-efficient of friction of the brake blocks and the rim, or pads and discs. Stopping distance can be estimated from the co-efficient of friction of the brakes, typically about 0.75. We take this as a percentage of gravity (G). So, $0.75 \times 9.81 = 7.35 \text{ m/s}^2$. Taking the bicycle travelling at 20 kph will have the following stopping distance. Initial velocity (U) = 5.6 m/s, final velocity (V) = 0, rate of deceleration (a) −7.35 and s is stopping distance

$$V^2 = U^2 + 2as$$

$$0^2 = 5.6^2 + 2 \times (-7.35) \times s$$

$$s = 2.13 \text{ m}.$$

Tech note

A car travelling at 40 m/s will take 109 m to stop on a dry road.

KINETIC ENERGY

Kinetic energy (KE) is the energy possessed by a moving body of mass (m) and velocity (v) such that:

$$KE = \frac{1}{2} mv^2$$

So, the bicycle and rider of mass 85 kg travelling at 5.6 m/s (20 kph or 12 mph) has a KE

$$KE = \frac{1}{2} 85 \times 5.6^2$$

$$= 1333 \text{ J} (1.3 \text{ kJ})$$

As the KE increases by the square of the velocity, an object doubling its velocity will have its KE multiplied by four.

This is the amount of work that the brakes will need to do to stop the bicycle and rider. The brakes convert this KE into heat, which is dissipated by rims or discs into the atmosphere.

Tech note

A small car of 1000 kg travelling at 50 kph (30 mph) will have a KE of 96,605 J or 96.6 kJ.

FRICTION

The co-efficient of friction μ (the Greek letter MU) is the ratio of the frictional force (F) required to slide a body over another body, divided by the force the normal force (N)

$$\mu = F / N$$

μ will always be less than unity (1). The braking force will depend on the following two points:

- The coefficient of friction of the brake blocks/pads against the rim/disc.
- The co-efficient of friction the tyres and road.

Both of these factors are also liable to variations with velocity and moisture.

Table 5.1 Typical coefficients of friction between tyre and road for different road surfaces

Road surface	Coefficient of friction (also called adhesion)
Dry concrete or asphalt	0.8–0 .9
Wet concrete or asphalt	0.4–0.7
Rolled gravel	0.6–0.7
Sand	0.3–0.4
Ice	0.1–0.2

Table 5.2 Typical coefficients of friction of brake blocks and rim

Material	Coefficient of friction dry	Coefficient of friction wet
Rubber block	0.95	0.05
Cork block	0.63	0.19

MECHNICAL ADVANTAGE

Mechanical brakes are designed to get the best mechanical advantage, that is, convert the pull on the brake lever to the highest possible force on the blocks or pads. This is done at both the lever and caliper ends. The different brake designs are mostly attempt to improve the mechanical advantage ratio.

Looking at a typical brake set up:

The lever is 12 cm long from the pivot and the point where the cable is attached is 2 cm from the pivot. This is 12/2 which gives a ratio of 6. The point where the other end of the cable is attached to the caliper is 8 cm from the pivot point, the block is 5 cm from the pivot this is 8/5, which gives a ratio of 1.6. Multiply the two ratios give 9.6. So a 30 N grip on the lever will give the block a force of 288 N against the rim.

Figure 5.1 Dual pivot front brake.

Figure 5.2 Dual pivot rear brake.

PASCAL'S LAW

Hydraulic brakes work on the basis of Pascal's Law. That is when a liquid is pressurised, the pressure is equal in all directions at 90° (normal) to the surface.

Tech note

The story of Pascal's Law is entertaining. French wine makers couldn't work out why when they hammered the cork into the neck of a bottle of wine occasionally, the bottom of the bottle would break off. They employed Blaise Pascal to investigate the problem. In about 1647, he find out that putting pressure anywhere on a fluid would increase the pressure in all the rest of it. The solution was to give wine bottle the dimpled bottoms, which are used now, so that the pressure was not going to force the bottom off.

HYDRAULIC ADVANTAGE

Following on from Pascal's Law, if you look at a hydraulic brake set up, you will see that the diameter of the piston in the cylinder operated by the lever is much smaller than the piston that moves pads in the caliper. If the brake lever is pulled such that it exerts a force of 20 N on the piston, and the piston is 1 cm^2 in area, the pressure in the fluid will be 20 N/cm^2. As Pascal's Law states that the pressure is equal in all direction, there will be a 20 N/cm^2 force against the piston in the caliper. If the area of the piston in the caliper is 3 cm^2 then the force on the piston will be 3×20 N/cm^2 that is 60 N.

Given a lever to pivot ratio of 6 as in the mechanical brakes, the force is multiplied by 6, this is multiplied by the hydraulic advantage of 3 giving an overall figure of 18. That is 18 times the force applied by the hand is applied to the hydraulic piston in the caliper. As can be seen, advantages can be much larger with hydraulic systems than mechanical ones.

Figure 5.3 Hydraulic hose for front brake in fork.

Figure 5.4 Detail of hydraulic hose at lower end of fork.

CABLE VERSUS HYDRAULIC

All cars have hydraulic braking systems, as do trucks and railway cars. Bicycles have retained cables up until recently for low cost and simplicity. Now, with some bicycles costing more than a cheap car, hydraulic brake systems along with electronic gears are becoming common. The following points are worth noting:

- Inner part of brake cables moves from inside to out, if the cable goes through a tight curve, the friction inside will increase proportionally. Cables can also rust and stick. Their mechanical advantage depends on the positions of the actuating mechanism, this can take a lot of space.
- Hydraulic fluid does not flow in the fluid line, it is simply displaced and subjected to pressure. The hydraulic advantage does not need a mechanical mechanism. Its main disadvantage is that it can leak if damaged.

Let's get comfortable— saddles and handle bars

Whatever you are riding, wherever you are riding, whenever you are riding, you need to be comfortable. What's more, your riding position will change as your body changes. You can get measured and fitted on to your bicycle, like a nice tailored suit, but even the best fitting suit might need the belt adjusting after a big dinner. So be prepared to adjust your setting to get comfortable.

Figure 6.1 Cycling club riders warming up before an event.

HUMAN POWER

The human body is like a form of engine. You feed it hydrocarbons and it gives out power. A good analogy to start with is a car engine. We put in fuel, it is burnt in the engine at about 2,000°C and as the gas expands, it pushes down the piston. This does useful work. A petrol engine is about 30% efficient, that is, about 30% of the energy from the petrol actually does useful work—the remainder warms up the engine and reduces the exhaust. A diesel engine is about 40% efficient; the human body is typically 25% efficient in its use of the energy in food.

The human body at rest is converting the energy, this is, called the Basal Metabolic Rate (BMR).

Table 6.1 Basal metabolic rates of organs at rest

Organ	Power consumed at rest—watts	Oxygen consumption at rest—millilitres per minute	Percentage of BMR
Brain	16	47	19
Heart	6	17	7
Kidney	9	26	10
Liver/spleen	23	67	27
Others	16	48	19
Skeleton muscle	15	45	18
Total	85	250	100

So, lying down asleep, the average person is using about 85 watts. Which leads us to functional threshold power—FTP, that is the amount of power output that a cyclist can sustain for one hour. Various sources show the range for active cyclists is between 100 and 400 watts, both the mean and mode are 250 watts.

If you watch cycle races, you will notice that the best climbers and, hence, winners of stage races tend to be of small stature—not always, I must add. The logic behind this is the concept of FTP in terms of watts per kilogram. In terms of racing car, we call it power to weight ratio, which is expressed in bhp per ton. Using the same group of cyclists that gives an average of 250 watts FTP, the watts per kilogram output for active cycling males' averages about 3.5 w/kg. To finish on the podium of a World Tour event, your output needs to be in excess of about 6 w/kg.

Figure 6.2 Cleats on MTB shoe.

BIOMECHANICS AND ERGONOMICS

Bicycles are very much for the individual, position is all, but this varies with rider comfort, required outcome, aerodynamics and what the rider enjoys.

Two factors affect bicycle rider's comfort and position, these are as follows:

- Frame configuration.
- Setting up of components—saddle, handle-bars.

You also need to know how you are going to ride it, touring, commuting, time-trialing, road-racing, off-road, MTB, BMX or any other form. The choice of frame will be the first thing, and the size of frame is the starting point.

Off-the-shelf framesets, that is, frame and forks, tend to be made in only a limited number of sizes. Custom-made framesets can be made to your exact requirements.

Road bicycle framesets, whether bought separately or as a complete machine tend to come in three sizes, these are small, medium and large. They may be labelled as such or given a numerical size in either inches or centimetres. However, you are likely to find that, the same sizes from different makers are in actual fact very different. Rather like designer fashion in clothing, frames with sloping top tubes confuse the sizing even more.

> **Mountain–MTB**–bicycle framesets are also labelled in the same way. The small, medium and large designations are usually accurate and similar, but the numerical labelling can be confusing. For example, I have a road bicycle that is designated 22 inch and an MTB that is designated 26 inch, both from the same manufacturer and both fit me the same.
>
> **Hybrid and BMX** bicycle frame sets do not conform to any of the sizing rules at all.

Manufactures of high-end, or elite, framesets usually publish the lengths of each tube component so that you can workout which frame will suit you the best.

Tech note

Probably, the most important point to remember is that one position will not suit you for all your riding, you'll either need to regularly re-configure settings or have several machines, unless you just do one kind of riding.

STANDOVER HEIGHT

The overall size of the frame is the best starting point, make sure that you can get on it comfortably. A good starting point is your inside leg measurement–also called your inner-seam length. More simply, the leg length of your trousers–this is simple for men, but women's trousers are not so simply labelled. Take your inside leg measurement and subtract 10 inches or 25 cm.

Table 6.2 Inside leg to frame size—the figures are rounded off

Inside leg inches	Inside leg cm	Frame size inches	Frame size cm	
30	76	20	50	Small
32	80	22	56	Medium
34	86	24	60	Large

The bottom bracket heights vary according to the intended use of the frameset, this effects the relationship between the length of the seat tube, which gives the frame, its generic size and the standover height. Sloping style top tubes also effect this measurement. With the bicycle on its intended wheel, you should be able to stand over the frame, wearing normal cycling shoes with about a 1 inch (2.5 cm) clearance. It is better to have a slightly smaller size than one which is too large. The smaller one is safer to ride, especially in traffic that requires stopping and starting.

SADDLE POSITION

Getting the saddle height correct is a major contributor to comfort, so is the fore/aft position. One of the knock-on effects from a badly positioned saddle is that the pedalling action is inefficient. The ball of the foot, where the toes pivot from, should be at the centre, the pivot point, of the pedal. Pedals with cleat fittings, or toe-clips, by their design hold the foot correctly.

To get the saddle height correct, get on the bicycle, either leaning against a wall, or being held by a colleague, sit in your normal comfortable position, then straighten your leg with your heel on the pedal. Adjust the seat height until when both legs, in turn, can be held straight—not stretching—with your heel on the pedal. Most people have slight variations in leg length, so make sure that you are comfortable on both sides. The rocking action of your ankles will accommodate the slight variations in your anatomy.

The fore/aft position of your saddle depends on the angle of leaning of your body and handle bar reach. It also depends on the saddle design. Roadster type bicycles and commuter/hybrids often come with a wider saddle, which gives a single seating position. In this case, it must be adjusted fore/aft to be comfortable for the single riding position. For club style riding on sports/elite bicycles, the rider will change positions from having the hands sitting on the tops of the bars, on the hoods and down on the hooks. This will change body angle. For this type of riding, the contact point with the saddle will change as the riding position changes. A long thin saddle allows this. To get this right, several different journeys will be needed, noting mentally the different positions and re-setting the saddle to suit. It's normal to actually carry a suitable Allen key so that adjustments can be made during the day. Mark the positions with a pencil or marker pen if possible, be prepared to go through changes back to the beginning—see the Tech Note section relating to tyre pressures too.

Saddle angle—the angle of the saddle can be altered to suit your position, this will affect your tendency to move backwards or forwards. For most riders, a perfectly level position is the best.

Personally, I start the season with a soft riding position—on my fixed gear bicycle, then move to a harder, more stretched position, when the season is underway on the geared machine.

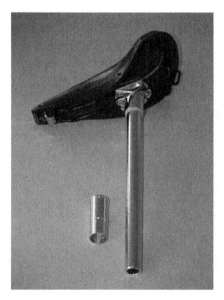

Figure 6.3 Seat post and size adaptor sleeve.

HANDLE BARS AND STEM

There is a great variety of handle bars and stems. The type of handle bars will depend on your type of riding; there are hundreds of different options in size and shape. Stems come in two main types—quill pattern and A-Head. The quill pattern allows the raising and lowering of the handle bars using one Allen key. The A-Head type has limited vertical adjustment depending on the number of spacers that are available. Both types are available in different lengths and a variety of angles. The vertical stem position allows different body angles. The length of the stem can be used to compensate for top tube length. Many frames are built to square proportions, that is, the top tube is the same length as the seat tube. This is ok if your legs, torso and arms are of the same size, but few people are actually in this category. So, choose a stem length to suit your body proportions. To do this, once you have got your saddle right, find your comfortable position leaning forwards. Typically, this is leaning your body forwards at 45 degrees to the horizontal with your upper arms at 90 degrees to your body. Then move your fore arm and hands to their natural position, this is where your handle bars should be, and shows the stem length needed. On a touring bicycle, this should be a position similar to that of playing a piano.

Tech note

Most bicycle manufacturers offerings for ladies' bicycles constitute loop frame shoppers—they seem to ignore the large number of lady racing, and sports, cyclists. Handle bars, stems and saddles are available to make lady cyclists of all sizes comfortable. These can be added to any frame.

Figure 6.4 **Aerodynamic tri-bars.**

Figure 6.5 **Multi-position loop bars - 1.**

Figure 6.6 Multi-position loop bars - 2.

Figure 6.7 Combined gear and brake lever.

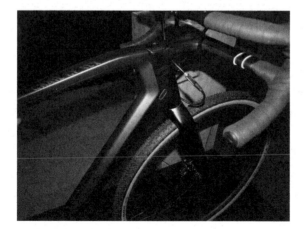

Figure 6.8 Dual bars - 1.

Figure 6.9 Dual bars - 2.

Figure 6.10 Comfortable bars and brake levers.

COMMUTING

It depends on your riding style, how long is your journey to work and what your actual job is. The author knows one rider who does an 85 mile round commute to work and back every day—on fixed gear too. Cycling to work is really good for you, as it makes your blood circulate and clears your mind. The watch word is comfort. Saddle bag for essentials, mudguards and a fairly upright position with suitable gearing and good lights.

ENDURANCE RIDING

Endurance riding is getting much more popular and competitive too. You know the conversations, *I rode to Scotland for my holidays*. The other person says, *I rode across Africa*.

There are a number of websites/blogs/social media where long-distance riders communicate and exchange house swaps and other accommodation information.

The types of bicycles chosen seems to be across the full range, but MTB would be an obvious choice; but I don't think ultra-endurance cyclists necessarily make obvious choices.

It's got to be fun, so you can be silly in your choice of machine. However, it would be disastrous if you couldn't get spare tyres.

CRANKS

The longer the crank in effect, the greater the leverage that you can apply, same as using a tyre lever or crow-bar. The longer it is, the more the turning effort—torque—you can apply. Longer cranks are preferred for hill climbing as its effect lowers the gear ratio—they also have the psychological effect of making the rider feel stronger.

Longer cranks are also more comfortable for riders with longer feet—they allow a more relaxed muscle movement of the ankle. Whereas riders with shorter feet may have a faster pedalling rate. There is also reverences in the various cycle training literature to a ratio between leg length and crank length. However, people with longer feet tend to have longer legs, so the rule of longer feet use longer cranks applies. More important, to both the casual and serious cyclist alike is pedalling style.

Pedalling style is something which is often over looked by riders. If the ball of the foot, that is where the toes pivot, is central over the pedal spindle, the angle can be used to both increase the cadence—pedalling rate—and give more effort onto the crank.

HOW TO ADJUST YOUR POSITION

Your ideal position will change on a regular basis. As your body weight changes and your muscle strength changes with the seasons, you will need to adjust your position. You can easily change your saddle and handle bars, to suit your personal preferences fairly quickly. Just changing the handle bar, tape can make the bicycle feel completely different. Just experiment to get yourself comfortable.

Try moving the saddle up and down a fraction, move it fore and aft, change the saddle angle. Just giving yourself wiggle room.

Handle bar position, propping yourself against a wall, bend your back up and down and move your arms until they feel comfortable—do you need to change the stem, longer, shorter or on with a slight angle to take the bars up or down.

Most off-the-peg cycles come with fairly standard Maes bend handle bars. These are flat on the top with a pretty standard hook shape. Maes bends are available in a variety of sizes, but you might find South of France bars—gently sloping bars, or North Road bars—flatter and more curvy, more comfortable.

A number of companies offer a cycle fitting service using jigs, but they can only suggest positions for you to try.

Chapter 7

Electrical power

Figure 7.1 Electric sport bicycle.

RULES AND REGULATIONS

The rules relating to the use of electric bicycles varies between both countries and types of bicycle. Looking at the UK for example, you can ride an electric bicycle in England, Scotland and Wales, if you are more than

85

14 years old and it meets certain requirements. The electric bicycles are defined as **Electrically Assisted Pedal Cycles—EAPC** in the UK Government rules.

By definition, an EAPC must have pedals that can be used to propel it. It must show either:

- the power output;
- the manufacturer of the motor.

It must also show either:

- the battery's voltage;
- the maximum speed of the bicycle.

Its electric motor:

- must have a maximum power output of 250 watts;
- should not be able to propel the bike when it's travelling more than 15.5 mph.

An EAPC can have more than 2 wheels (for example, a tricycle).

Where you can ride, if a bicycle meets the EAPC requirements, it's classed as a normal bicycle. This means you can ride it on cycle paths and anywhere else bicycles are allowed.

Any other electrically propelled bicycle that does not meet EAPC rules is classed as a motorcycle or moped and needs to be registered and taxed. You need a driving licence and must wear a crash helmet to ride it. This electric bicycle must be type approved if it does not meet EAPC rules or can be propelled without pedalling—such as 'twist and go'.

Be aware that there is a large number of 'home-made' non-EAPC electric bicycles in use. These may have been built for off-road use in the same way as Moto-X machines. The use of these is limited to restricted venues.

Still in the UK, but with different rules, in Northern Ireland electric bicycles must be registered, taxed and insured. In addition, the rider must have a moped licence.

Other countries

In France, an electric bicycle is one which does less 25 Km/h (15.5 mph). EU regulations suggest that insurance is essential. In the United States, the regulations vary state by state; some states have very high cyclist casualty rates, probably attributable to high car density and the misuse of narcotics, which are readily available in these areas.

Figure 7.2 Battery in down tube.

ELECTRIC BICYCLE CONSTRUCTION

The construction of the current range of electric bicycles is based on the construction of conventional, non-electric bicycles with the addition of three main components and a control switch/display panel. The three components are **motor, battery and sensor**. We'll discuss each in turn.

Tech note

Watt is the unit of power, which has been in use since we started to use electricity. It is the product of multiplying the voltage by the amperage. In other words:

Volts × Amps = Watts.

If you wish to compare it to your car engine, 746 watts equals to 1 Horse Power. Other ways of saying of Horse Power (HP) are Cheval Vapour in French (CV); and in German Pferde Stracker (PS).

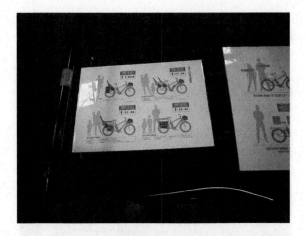

Figure 7.3 Optional configurations of electric bicycles.

Motor

The motor provides the tractive force to drive the bicycle. On an EAPC machine, the maximum allowed power output is 250 watts. That is about one-third of a horse power. To put this into context, an average healthy male can maintain this power-output for about an hour. A racing cyclist will maintain more than double of this for several hours. The power output of the motors used by the main cycle manufactures tend to be rated slightly below the 250 watts maximum, typically, this is 230 watts. The power output is only part of the design specification to be balanced with size, weight and aesthetics.

Also, to be borne in mind is that power can be rated in three different ways:

- **Maximum power:** the maximum amount that can be developed at a particular speed.
- **Continuous power:** this should be identified as a particular output at a particular speed for a set time.
- **RMS value:** the calculated average value. As a general rule:

RMS Power Value = Maximum Power × 0.7071

So, for a 250 watts maximum motor, the RMS calculation is:

RMS Power Value = 250 watts × 0.7071

176.775 watts.

To make this even more complex, electric motors are often rated on their electricity consumption not their output. The efficiency of electric motors varies considerably on their construction. Typically, the efficiency of an electric motor may be 85%, that is, 85% of the electrical power actually does useful work. So, an 85% efficient electric motor rated at 250 watts will give:

$$250 \text{ watts} \times 85 / 100 = 212.5 \text{ watts}.$$

There are three commonly used positions for the electric motor, these are as follows.

Front wheel motors

- Probably the cheapest option as the motor fits simply in the wheel and does not impact on the gears.
- It will accommodate a reasonable level of power, ideal for city/ town use.
- Good for providing steady assistance for hill climbing.
- May make the steering heavy and the machine seem a little more difficult to manoeuvre.
- As only part of the load of the machine is on the front wheel, there may be problems with traction in icy or wet conditions.

Figure 7.4 Step through bicycle with electric motor in front wheel.

Figure 7.5 Electric motor control panel.

Rear wheel motors

- Offers both better traction and handling as the drive is as it is on a normal bicycle or motorcycle.
- As the natural feel of bicycle with a normal drive line.
- Stealth appearance, just looks like a large hub gear.

Centre or bottom-bracket mounted motors

- Provides a higher level of assistance as larger motors can be used and the drive is through the gear system.
- This arrangement is better for long steep hills.
- More sensitive to your riding style, so providing the power exactly when needed by the rider.

Figure 7.6 Electric motor in bottom bracket – 1.

Figure 7.7 Electric motor in bottom bracket – 2.

Battery: This stores the electrical power in chemical form. The batteries are similar to those used in industrial standard power tools. They generally work at either 24, 36 or 48 volt. The cell construction is usually lithium-ion (Li-ion); however, other types that are available include lead-acid (SLA), nickel-cadmium (NiCd), nickel-metal hydride (NiMh) and lithium-ion polymer (Li-pol). Li-ion batteries have the advantage of not having a memory—you do not need to fully discharge them before re-charging, however, they are fairly expensive to make. Li-ion type battery is used in mobile telephones, laptop computers and other similar devices.

There are three basic **battery rating**:

- The nominal voltage at which they work on electric bicycles, these are generally either 24, 36, or 48 volt. However, be aware that these are only nominal voltages—in other words approximate voltages. The voltage will vary with the state of charge of the battery and the load applied. When a load, for instance the motor, is applied with a voltmeter attached across the battery terminals, you will see a drop in the reading. This is called volt-drop and is due to the internal resistance of the battery. A faulty/old battery will have a greater volt-drop than a new one.
- The Amperehour rating—Ah. This is not as straight forward as you would expect. A typical electric bicycle battery may be rated at 10 Ah, this does not mean that it will give out 10 Amps for 1 hour, nor indeed 1 Amp for 10 hours. Batteries are tested in several different ways and averaging calculations are made.
- Watthours—Wh. This may be a fairer rating as watts are the product of volts times amps. Typically, the batteries used on electric bicycles are about 500 Wh. However, do not expect the battery to give out 500 watts for one hour.

SAFETY NOTE

There has been a great growth in battery sales, as a battery is a store of energy in chemical form, it MUST be treated with care. New European standards are coming in to use—EN 50604.

Care of batteries: To keep your battery safe and maintain a long life, the following points should be noted.

- Store it in a cool dry place.
- Keep away from children—secure locked cupboard.
- Keep fully charged, but not overcharged.
- Have the operating instructions to hand—read them before using or charging the battery.
- Connect in the prescribed way to the electric bicycle, ensuring that the cable and connections are firm and secure.
- Dispose of used batteries following manufacturer's and local environmental advice.

Figure 7.8 Close up of front wheel electric motor.

Sensor: This is the component that switches on the motor when you start pedalling. There are two sorts of sensors:

- **Speed sensor:** This is the cheapest of the two, it switches on the motor when it senses movement and increases with rider cadence (pedalling rate). Ideal for commuter electric bicycles.
- **Torque sensor:** This increases the motor output as more pedalling effort (torque) is applied. Riding with one makes any ordinary cyclist feel like a Tour de France winner. There are several different types of torque sensors— bottom bracket, chain, crank and derailleur hanger. They all work on the principle of measuring the torque applied by the rider—not the speed.

Chapter 8

Add-ons and kit

LIGHTING

The UK Government regulations on bicycle lighting are following.

Any cycle which is used between sunset and sunrise must be fitted with the following:

- *white front light;*
- *red rear light;*
- *red rear reflector;*
- *amber/yellow pedal reflectors—front and rear on each pedal.*

The lamps may be steady or flashing, or a mixture, for example, steady at the front and flashing at the rear. A steady light is recommended at the front, when the cycle is used in areas without good street lighting.

If either of the lights is capable of emitting a steady light, then it must conform to BS 61023 and be marked accordingly, even if used in flashing mode.

Purely flashing lights are not required to conform to BS61023, but the flash rate must be between 60 and 240 equal flashes per minute (1 to 4 per second) and the luminous intensity must be at least 4 candela (This should be advised by the manufacturer). The pedal reflectors and rear reflector must conform to BS 61022.

Lights and reflectors not conforming to the BS, but conforming to a corresponding standard of another EC country and marked accordingly, are considered to comply as long as that standard provides an equivalent level of safety.

Lights are not required to be fitted on a bicycle at the point of sale—but if they are fitted, then they must comply with these regulations.

Optional lamps and reflectors

Additional lighting to the above-mentioned obligatory lights is permitted under certain conditions:

- *it must not dazzle other road users*
- *it must be the correct colour (white to front, red to rear)*

- *if it flashes, it must conform to the required flash rate (1 to 4 equal flashes per second)*

Optional lights are not required to conform to BS 61023 and there is no minimum level of intensity. So, for example, on the rear of the cycle, a cyclist may wish to have both a steady red lamp, which conforms to BS 61023, and an additional flashing lamp, which is not meeting the minimum level of 4 candela.

(www.uk.gov/government/publications/pedal-cycles-lighting as of February 2020) You are advised to check for updates and local regulations.

If you analyse the law, it is about see and be seen, very sensible. The law has been changed over the years to incorporate new technology. The time-trial organising body suggests the use of rear lights during events on public roads; these may be steady or flashing. The use of two rear lights, one steady and one flashing is perhaps the best. A similar situation applies for the front—perhaps using high-power front lights for serious winter training.

Nowadays, there is also a trend to fit amber lights to frame and/or forks to make riders visible from the side—a particular problem in areas of pooror non-existent street lighting.

Figure 8.1 **Add-ons for gravel adventure bicycle.**

COMPUTERS

There are two types of bicycle computers:

- **Dead reckoning ones:** These measure the distance covered by the wheel against time. They must be set to wheel size and the time must set like an ordinary watch. The wheel size setting relates to the actual distance covered by the wheel in one revolution—usually in millimetres.

- The other is GPS based, so distance and climbing height can be measured against the standard mapping. Time is set from the internet or radio clock signal.

The GPS—in full called the Global Positioning System—is used by a number of hardware manufacturers and software/firmware companies. The two that are popular in the UK and most of Europe and the USA are Strada and GPS.

Many cycling clubs run league tables based on Strada. Club members upload their training and other journeys to the club website so that there are lists of data such as:

- Distance covered
- Maximum speed
- Times/speeds for sections of roads—in particular, for hills

Strada has led to a sub-culture of riders trying to beat each other and comparing data. Beware, some cheat, on my local hill route a rider posted an absolutely shattering time, later we discovered that he had done it on his motorcycle.

Figure 8.2 **Allows you to train without going outside.**

ACCESSSORIES

This is gigantic business, it includes:

- Add-on items
- Luggage

- Upgrade equipment
- Lighting
- Tools
- Clothing

A main supplier of accessories is Extra (UK) Ltd—their current catalogue has 356 pages.

CLOTHING

Clothing for cycling has been, and still is, an issue for cyclist.

What shall I wear?

The current trend club racing cyclists, and has been for several years, is the MAMIL look—middle-aged male in Lycra. This also applies ladies too. Lycra is looked upon as a sort of miracle fabric, warm, quick-drying and smart in appearance. The current versions used by *World Tour* teams are aerodynamic too.

There are many alternatives, plus-two trousers for touring in winter are very comfortable. For off-road riding, especially if it includes off-bike activities such as camping, then cotton semi-baggy shorts are preferable.

For those riders who like vintage/Eroica type events wool tops and shorts are readily available.

Figure 8.3 Helmet with built-in rear light.

BOTTLES/BIDONS

It is sensible to keep your body hydrated when cycling, so some sort of drinks bottle is very useful. The word bidon is used for cycling bottles—it is simply a French word to describe a sealed drinks container, these were made popular by their use in the Tour de France. The riders tend to throw them on the roadside once empty, to make space for the next drink. Collecting bidons is a past-time for many racing fans, as many bidons are brightly coloured and carry logos of the sponsors—they also have a story to tell, who they belonged to and in which race. Rather like collecting autographs or football programmes.

SHOES/CLEATS

There is a wide variety of shoe types depending on usage, such as:

- Elite road shoes
- Off-road shoes
- Road shoes
- Touring shoes
- Triathlon shoes

The shoes attach to the pedals with cleats, the two most popular systems are SPD and look.

For vintage/Eroica type events, toe-clips and shoes with shoe-plates are available.

LUGGAGE

There are many types of luggage or luggage systems. Such as:

- Rear panniers
- Front panniers
- Top tube mounted bags
- Handle bar bags
- Saddle bags

Keeping the luggage low keeps the centre of gravity down, making the bicycle more stable. Let the actual bicycle take the load, rather than using a rug/knapp sack on your person.

Chapter 9

Materials for bicycles

Bicycles tend to be made in a traditional way using traditional materials. This section discusses some of the materials that are used and their properties.

Engineers tend to classify materials into two major groups, each with two sub-groups. The major groups are metallic materials and non-metallic materials. We'll look at each in turn.

Table 9.1 Bicycle engineering materials

Bicycle engineering materials	
Metallic materials	
Ferrous—contains iron	Non-Ferrous—does not contain iron
Iron in various forms	Aluminium
Low carbon steel	Brass—copper & zinc
Medium carbon steel	Bronze—copper & tin
High carbon steel	Chromium
Alloy steel	Copper
	Titanium
Non-metallic materials	
Natural—occur in nature	Synthetic—man-made materials
Leather	Carbon fibre
Wood	GRP—glass fibre
Wool	Lycra/Spandex/elastane
Bamboo	Thermo-plastics
Cotton cloth	Thermo-setting plastics

Metallic materials

The metallic group is divided into two sub-groups, these are **ferrous metals** and **non-ferrous metals**. Ferrous simply means iron, all ferrous metals contain iron. Non-ferrous metals do not contain iron.

Iron is dug from the ground and heated in a furnace—there are several different types of furnaces—and mixed with carbon to form steel. Steel has been a popular choice for bicycle construction since the start of cycling. Steel was extensively used long before bicycles were invented. When we talk about steel, it is important to realise that there are several major categories of steel: low carbon, medium carbon, high carbon and many types of alloy steel. When we talk about alloy steel, we simply mean that it is steel mixed with another element.

Tech note

Alloy: A mixture of a metal and another element.

MANUFACTURE OF STEEL FOR BICYCLES

Iron ore, which is dug up from the ground, is fed into a blast furnace together with limestone and coke. The coke is used as a source of heat and the limestone as a flux, which is an agent that cleans and helps the flow of the metal. It separates the metal from the impurities in the mixture. The molten metal is now poured out of the furnace into moulds to form what are called pigs—chunks of iron which resemble to shape of a pig's body. Because of the burning process, the pig iron contains between 3 and 4 percent carbon.

The pig iron is now changed into steel by re-heating in a furnace and blasting with air to reduce the carbon content to between 0.08 and 0.20 percent. The term blast furnace is used, though there are other types of processes.

Casting

The steel is cast into ingots or into a continuous rolling slab depending on the process and purpose of the material.

Pickling

The next stage is to remove the black scale from the surface of the metal, this is a process called pickling—the steel is run through a bath, or shower, of either hydrochloric acid or sulphuric acid. This ensures that the surface of the steel is clean.

Cold rolling and hot rolling

To make sheet steel, which may be converted into tubing by bending and joining, the ingots, or continuous slab may be rolled either when it is hot or when it is cold. Rolling changes the structure of the steel and needs to be followed by annealing and tempering stages. Tubing made from sheet steel is used in the construction of some utility bicycles, and the conduit tubing and trunking are used in building construction, which carries only minimum stresses.

Annealing and tempering

Annealing is a method of treating the steel, taking away the internal stress. It needs to be stress free, or softened, to be able to be worked into the desired shape. Tempering is a process of making it the correct hardness. The annealing process usually means heating and cooling the steel in a controlled, oxygen free atmosphere. The tempering involves re-heating to a set temperature and cooling at a specific rate. As you can see this process uses a large amount of energy—the material has been heated and cooled four times: original casting, blasting, annealing and tempering.

Hot drawing and cold drawing

Sheet steel can be bent into tubes and seam welded. The problem is the seam is weak and may tear apart under stress. Therefore, tubing made in this way tends to be made thicker to give the required factor of safety.

Tech note

Factor of safety is the number of times that the maximum load, that a component can carry, is divided by the expected load; this is expressed as a ratio or percentage. If a bicycle tube can carry a load of 650 kg before breaking, and the load is 65 kg (weight of a typical rider) then the calculation is 650 / 65 = 10. The factor of safety is 10. This calculation is now being used in the design of racing bicycles to get the maximum lightness. To be borne in mind is that dynamic stress is greater than static stress for the same load generally called 2 σ (2 Sigma).

Reynolds in the UK and Columbus in Italy developed ways of drawing tubing without seams, seamless tubing. Being seamless, the tubing is equally strong across entirety.

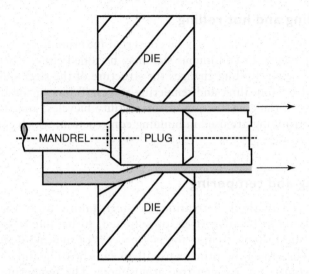

Figure 9.1 Drawing steel tube.

The hot ingot is held in a die with a plug, the tube is formed by being pulled over the plug. This process may be carried out several times to get the required tube wall thickness.

CLASSIFICATIONS OF STEEL

Steel is a highly developed product and is classified in a number of ways, general classifications are:

- Cold forming steels
- Carbon steels
- Alloy steels
- Free cutting steels
- Spring steels
- Rust-resisting and stainless steels

Cold forming steels: This is in effect sheet steel, used on pressed steel bicycle components, mainly in the utility cycle sector. Components such as brake levers and pedals are made from pressed steel as they can be made in a very large numbers very quickly. Using automated presses, a brake lever or pedal can be made in less than five seconds—that is, up to 18,000 per day. Just to put this into perspective, one UK cycle chain store sells 26,000 bicycles each week and the number is rising. The process for the manufacture of a component such as a brake lever requires two stages. First the material is cut to size—that is, it is stamped out like you might cut a biscuit out of pastry. Then it is moved to a machine which will bend it, also called forming, into shape against a die. The cutters for the shape and the dies for the forming can be changed in these machines in minutes so that a factory can make brake levers in the morning and pedals in the afternoon—or any other similar product. Such companies tend to supply pressed products to a range of different industries, such as bicycle manufacturers, the automotive industry and the construction industry.

SAFETY NOTE

Some bicycle components may look and feel like they are manufactured from sheet steel, but they may not behave in a way which you would expect. For example, manufacturers in the Far East often use different materials and processes to those in Europe and the USA, so welding and other repair or modification procedures may not be possible.

Carbon steels: These are used for a large number of small bicycle components, such as chains, gear blocks, axles and bearings and traditional steel frames. There is a large variation in carbon steels, this leads to a set of general classifications of carbon steel:

- Low carbon steel—also called mild steel—0.10–0.25% carbon
- Medium carbon steel—0.20–0.50% carbon
- High carbon steel—0.50–2.00% carbon
- Tool steel
- Micro-alloyed steel

Low carbon steel is soft, ductile and malleable and, therefore, can be easily formed into shape. It cannot be hardened and tempered by heating and quenching; but it can be case-hardened and it will work harden. Case-hardening is a process of coating the surface of the steel component with a high carbon content chemical and heating to a set temperature. When the component cools, the surface is hard like high carbon steel and the underside remains soft and malleable. This process is used on hub bearing surfaces; if you look at a hub cone closely, you will be able to see the different colours of the metal. The advantages of this are that the axle and cones can be made of low carbon steel, which is both easier to machine and cheaper to buy and then giving a wear resistant surface for the bearing.

Low carbon steel is usually sufficiently strong for many components on bicycles, it is also cheap and plentiful, most steel suppliers can offer this readily off-the-shelf.

Medium carbon steel is much tougher and not as easy to bend or machine; it can be hardened and tempered.

Tech note

You will find components made from carbon steel at different price points, be aware of marketing scams—for example, the most expensive chain may not be made from the most suitable material, it may just be because it is a particular colour.

ALLOYING METALS USED WITH STEEL IN BICYCLE FRAMES

Chrome: A lustrous, brittle hard metal used to add corrosion resistance. It is the main additive in stainless steel. Abbreviation is **Cr**.

Manganese: Used in stainless steel to resist corrosion. Increases hardenability and tensile strength. Abbreviation is **Mn**.

Molybdenum: Used to enhance strength, improve the hardenability and weldability properties and add toughness. It also improves corrosion resistance and high-temperature deformation. Abbreviation is **Mo**.

Vanadium: Gives added resistance to corrosion and resistance to acids and alkalis. Abbreviation is **V**.

RESEARCH

If you are interested in researching into the materials and the manufacturing of bicycles, you'll find the following two organizations helpful:

Advanced Manufacturing Research Centre at the University of Sheffield.

Advanced Materials Research Group at the University of Nottingham.

It is the author's opinion that as bicycling is now in a new phase of development and open to new and innovative ideas, these may be large step changes, or small discreet developments.

Extrusion: Shaping components by forcing the metal through a shaped hole. The best way to explain this is to think about piping icing on to a cake. When you squeeze the icing bag, it comes out through a shaped end with a profile. Bicycle rims are extruded then curved and joined.

Hydroformed: Malleable metals such as aluminium can be formed into fairly complex shapes by hydroforming. This practice is extensively used on large diameter shaped frame tubes, those typically used on

MTB bicycles. The basic tube is put into a die assembly and then a liquid, either water based or oil based, is fed under pressure into the tube, forcing it outwards into the shape of the die.

Mar-aging steel: Also written **maraging steel** without the hyphen. This word is a combination of martensitic and aging. It is a process of adding toughness and strength to low-carbon ultra-high strength steels by heating to a high temperature for several hours before cooling. Ultra-high strength steels get their strength from intermetallic compounds, not added carbon. The compounds may include cobalt, molybdenum, titanium and niobium.

Stainless precipitation hardening steel: These are low carbon steels that have fairly high percentages of manganese, chromium, nickel, copper and titanium. The non-ferrous metals precipitate to make the steel hard. Precipitation means falling, a word that weather forecasters use for raining. In this case, it is the even distribution of these compounds of non-ferrous metals in the steel, like rain drops, that make the steel hard.

Cold worked: It means steel which is rolled out when it is cold. This changes the grain structure and so makes the metal harder and stronger, but reduces ductility.

Seamless: Tubing made from billet; not rolled and seam welded.

Air-hardening steel: This is fairly high carbon, 0.5–2% with the addition of molybdenum, chromium and manganese. It is hardened by heating to between about 800 and 900°C then cooling in air. The heating may be carried out in vacuum furnace.

Butting: Making the frame tubes thicker in places. Double butting is the most common; the tubes are butted at each end where they join the other tubes.

WORK HARDENING AND FATIGUE FAILURE

We talk about hardening as a good thing, but work hardening is different, it is a bad thing. Work hardening and fatigue failure lead to component breakage. Aluminium and copper both go hard due to time and vibrations. That is, they harden without being noticeably stress loaded. Steel does not do this.

If a steel frame is not over loaded, it will retain its strength for life time. For example, apart from the cycling world, the use of steel is in joists in buildings, these will remain straight and true, if the building is not over loaded.

An aluminium frame will work harden with time and normal road vibrations; therefore, it has a finite life span leading to eventual fatigue failure.

Welding and brazing changes the structure of the metals at the area of the joint, creating a point more susceptible to failure. The fracture usually occurs about 3 mm from the actual welded area, not of the actual weld itself.

From the author's experience, steel frames usually fail at the fork crown joint and where the rear stays are attached to the drop-outs. Aluminium frames are liable to fail at any point depending on construction methods and usage.

PROPERTIES OF MATERIALS

Stress: This is usually measured in mega-Pascals—MPa. The load in mega-Newtons—mN over the cross-sectional area in metres—m. There are several types of stress, metals are usually judged by their ultimate tensile stress—UTS. That is, the level of stress at which they will break. Making component of thicker metal will increase the load which it can carry for any given UTS. You may use the same type metal for two frames, but if it is going to be subjected to more loads, as in MTB/ATB, then you may want to use thicker tubes, hence the use of butting tubes at the ends.

Elongation: It is the amount by which something elongates, grows longer, compared to its original length. Also, the terms **deformation** and **extension** are used where it is not a simple change in length.

Strain: The ratio of elongation divided by the original length, usually expressed as a percentage.

Youngs Modulus: The ratio of stress divided by strain.

Strength: Usually refers to the UTS.

Factor of safety: The number of times that the maximum load is compared to the expected load.

Elastic limit: The stress at which a metal does not return to its original shape. Steel is up to a point elastic; you bend it and it bends back. Bend it more and it stays bent.

Stiffness: The load needed to bend a tube or stay.

Strength to weight: The UTS as a ratio of the density. UTS in MPa, density in gram/cm^3.

Stiffness to weight: The stiffness as a ratio to density.

Aging T numbers: A set of standards that aluminium alloy is hardened to. It is expressed as a working standard giving temperature that the metal is heated to, the length of time it is held at this temperature and the cooling process.

Welding and brazing dissimilar metals: The welding and brazing of dissimilar metals is possible with modern methods and fluxes. Of course, the joints and parts will have different strengths and properties to those of a normal, single metal joint. You should check with the material suppliers and carry out a test joint before using this in a real-life situation.

EN standards: European Standards, literally European Norm. British Standards (BS) have merged with these. In America, the equivalent is ANSI, Germany has DIN and Japan JIS. There are other equivalents used around the world. The number runs in to hundreds of variations.

TAKE CARE

When purchasing a frame, or tubing, take care to ensure that it is what it is advertised to be. Reproduction stickers and transfers are available for most frame and frame tube makers on websites. These stickers are often used for renovation purposes, but satisfy yourself as to the authenticity of the actual item before purchase.

REYNOLDS TUBING

Reynolds Tubing has been used for bicycle frames for over one hundred years; Reynolds invented the system of butting in 1898. It is made by Reynolds Technology Ltd., Shaftmoor Industrial Estate, Shaftmoor Lane, Birmingham, B28 8SP. They offer a range of readymade tube sets; they are named and labelled as to the metal alloy that they are made from.

953 Mar-aging stainless steel: It combines resilience with an extremely low weight. It has high-impact strength and fatigue resistance. It is ideal for racing bicycle frames.

921 cold worked stainless steel: High-strength, austenitic precision-welded stainless steel. This can be shaped by frame builders without further heat treatment. Because of the nature, it is suitable for a wide range of bicycles including utility, BMX, hybrid, ATB, 29er, XC and cargo bikes as well as road and touring forks.

853 seamless air hardening heat treated steel: Light in weight with increased strength after welding. 853 is heat treated to give high strength and damage resistance. Strong, durable and with excellent fatigue properties, ideal for heavy and stronger riders. Suitable for BMX, ATB and endurance bicycles.

725 heat treated chrome–molybdenum steel: Butted and heat-treated Cr–Mo steel. A thin walled tubing giving a weight advantage over non-heat-treated tubing. Can be TIG welded and combined in frame sets with 853 and 631 tubes.

631 seamless air hardened steel: Similar in chemistry to 853 and 631, it is cold worked and is air hardened after welding. Tough, durable and comfortable, 631 frames are particularly suitable for long distance riding, ATB and BMX. It can be used for both racing and touring forks. It can be both welded and fillet brazed.

531 manganese–molybdenum cold worked steel: This is the grandfather of all bicycle tube sets, it has been available in this format since 1935. It is perfect for brazed construction with lugs. It has a long history of use in bicycle frame construction. It is also used for racing motorcycle

frames, racing car sub-frames, aircraft spars and struts and the Trust 2 World Land Speed Record Car chassis.

525 cold worked chrome–molybdenum steel: This is a mandrel butted frame tube made to high accuracy. It is suitable for both high accuracy welding and fillet brazing. It has the dual advantages of being both light and competitively priced. It is suitable for most bicycle frame applications.

520 cold worked chrome–molybdenum steel: Similar to 525 in properties. It is manufactured in Taiwan under licence. It is very competitive in price and, therefore, suitable for large scale production.

7005 aluminium alloy: This is an industry standard aluminium alloy with zinc and magnesium elements. It is ideal for strong lightweight frame sets. The use of the T6 treatment process is recommended after frame construction. 7005 aluminium is also used for many applications outside the bicycle industry.

Tech note

T6 heat treatment condition—a procedure for ensuring that the welded aluminium frame is uniformly strong after welding. It is heated up to about 530°C for a period of time, then cooled in a water bath. There are a number of treatments with T numbers.

6061 aluminium alloy: Aluminium is alloyed with silicone and magnesium to provide a relatively low cost and very light weight material. It is very ductile and lends itself to the hydro-forming of the non-round shapes used for some bicycle frames. It is readily weldable and used for a wide range of applications outside the bicycle industry. It may be heat treated using the T6 process.

6-4 Ti seamless ELI grade titanium: Manufactured from custom made billet, it is the only mandrel butted seamless 6–4 tubing. It is 6 percent aluminium and 4 percent vanadium made to ELI grade standards. Very light weight, very durable and the highest fatigue resistance, so usable for high quality bicycle frames.

Tech note

ELI Titanium—extra low interstitial gas purity. This means that the titanium is reduced in oxygen and iron impurities, making it stronger and less liable to fracture. ELI Titanium is used for both medical purposes—such as joining fractured bones and aerospace engineering.

3-2.5 Ti seamless grade titanium: Three percent aluminium and 2.5 percent vanadium. Can be cold worked for custom designs and is easy to weld. So, it is ideal for custom built bicycles and special applications. More readily available than 6-4 ELI grade tubing.

Table 9.2 Reynold tubes strength and density (relative weight)

Tubing number	Material	UTS MPa	Density gram/cm³
953	Stainless Steel	1750–2050	7.8
921	Stainless Steel	950–1080	7.9
853	Steel	1200–1400	7.78
725	Chrome–Molybdenum Steel	1080–1280	7.78
631	Air Hardening Steel	800–900	7.78
531	Manganese–Molybdenum Steel	650–850	7.8
525	Chrome–Molybdenum Steel	700–850	7.78
520	Chrome–Molybdenum Steel	700–900	7.78
7005	Aluminium Alloy	400	2.78
6061	Aluminium Alloy	325	2.7
6-4Ti	Titanium	900–1150	4.42
3-2.5Ti	Titanium	810–960	4.48

RACER NOTE

Almost 25 percent of Tour de France racers have been won on bicycles made from Reynolds Tubing.

COLUMBUS TUBING

Columbus tubing is made in Milan, Italy. The company was founded by Angelo Luigi Colombo in 1919. It now constitutes two companies, these are Gilco and Trafiltubi. Gilco became well known for the tubing used for the chassis of Ferraris. The range of Columbus tubing currently available is:

Zonal 7000 aluminium alloy: This is an alloy of aluminium, zinc and magnesium. It is designated for amateur level road and MTB competition use. UTS is 420MPa.

Spirit alloy steel: A very high strength steel containing niobium. This is used for professional road racing and triathlon. UTS is 1050–1250MPa.

Life alloy steel: This is used for professional level MTB, triathlon and road frames.

Zona alloy steel: A non-symmetrical butted tubing for top of the range custom, competition road and MTB frames. It is alloyed chromium and molybdenum. UTS is 800MPa.

Megatubes: A range of oversize and specially shaped tubes in a range of materials.

Forks in cold drawn and cold shaped alloy steel: These use pilgrim-rolling technology. Pilgrim means that one undriven roller follows the driven roller. The alloy is chrome molybdenum.

Columbus XCr: Seamless stainless steel from billet. UTS 1250–1350MPa. A high-end professional frame material.

Columbus niobium: An alloy steel with manganese, nickel, molybdenum and niobium. UTS 1050–1150MPa.

Columbus 25CrMo4: A seamless alloy steel with high chromium content. UTS 800MPa.

OTHER TUBING SUPPLIERS

It should be noted that there are other bicycle frame tube suppliers. Reynolds and Columbus are the most popular brands in the club and sporting areas of bicycling. The author uses both brands as they offer reliability and regularity in both supply and performance.

Chapter 10

Soldering, brazing and welding

The joining of bicycle tubes to make a frame may be carried out by several different methods. The most common ones are soldering and brazing. For each of these general methods, there are several variations. This chapter discusses these methods to help enable the aspiring bicycle engineer to build, repair or identify a bicycle frame.

Tech note

Not all frame materials can be joined in the same way, this is further discussed in Chapter 9.

COMPARISON OF FUSION AND NON-FUSION JOINING PROCESSES

The joining of metals by processes employing a fusion of some kind, that is, the melting of metal. There are different types of fusion; they may be classified as follows:

Total fusion

Temperature range: 1130–1550°C approximately.
Processes: Oxy-acetylene welding, manual metal arc welding and inert-gas metal (IGM) arc welding. In other words, a welded frame usually without lugs.

Skin fusion

Temperature range: 620–950°C approximately.
Processes: Flame brazing, silver soldering, aluminium brazing and bronze welding. A lugged frame or one which is fillet brazed.

Surface fusion

Temperature range: 183–310°C approximately.
Process: soft soldering. Used on lightly loaded components, and for electrical and electronic components.

In total fusion, the parent metal and if used, the filler metal, are both completely melted during the jointing. Tubes can be fused together without additional filler metal being added. Oxy-acetylene welding and manual metal arc welding were the first processes to employ total fusion. In recent years, they have been supplemented by methods such as inert-gas arc welding, metal inert-gas/metal active gas (MIG/MAG) and tungsten inert-gas (TIG) welding, carbon dioxide welding and atomic hydrogen welding. Welding is normally carried out at high-temperature ranges, the actual temperature being the melting point of the particular metal which is being joined. The parent metal is totally melted throughout its thickness and in some cases, molten filler metal of the correct composition is added by means of rods or consumable electrodes of convenient size. A neat reinforcement weld bead is usually left protruding above the surface of the parent metal as this gives good mechanical proper ties in the completed weld. Most metals and alloys can be welded effectively, but there are certain exceptions which, because of their physical properties, are best joined by alternative methods.

In skin fusion, the skin or surface grain structure only of the parent metal is fused to allow the molten filler metal to form an alloy with the parent metal. Hard solders are used in this process, and as these have greater shear strength than the tensile strength, the tensile strength of the joint must be increased by increasing the total surface area between the metals. The simplest method of achieving this is by using a lapped joint in which the molten metal flows between the adjoining surfaces, this accounts for the use of lugs in joining frame tubes. The strength of the joint will be dependent upon the wetted area between the parts to be joined. Skin fusion has several advantages. First, since the filler metals used in these processes have melting points lower than the parent metal to which they are being applied, a lower level of heat is needed than in total fusion and in consequence, distortion is reduced. Second, dissimilar metals can be joined by applying the correct amount of heat to each parent metal, when the skins of both will form an alloy with the molten hard solder. Frame lugs are usually cast from a different alloy of steel to that of the frame tubes. Third, since only the skin of the parent metal is fused, a capillary gap is formed in the lap joint and the molten filler metal is drawn into the space between the surfaces of the metals; this allows easy assembly and wriggles room to adjust the frame angles before making the joint permanent. The filler material, also called spelter, will fill the space available when at the correct temperature to flow.

In surface fusion, the depth of penetration of the molten solder into the surfaces to be joined is so shallow that it forms an intermetallic layer which bonds the surfaces together. The process employs soft solders, which are composed mainly of lead and tin. As these also have a low resistance to a tensile pulling force, the joint design must be similar to that of the skin fusion process, i.e., a lapped joint.

SOFT AND HARD SOLDERS

Despite the growing use of welding, the technique of soldering remains one of the most familiar in the fabrication of sheet metal articles, and the art of soldering continues to occupy an important place in the workshop. While soldering is comparatively simple, it requires care and skill and can only be learnt by experience.

Soldering and brazing are methods of joining components by lapping them together and using a low-melting point alloy so that the parent material is not melted. Soldering as a means of joining metal sheets has the advantage of simplicity in apparatus and manipulation, and with suitable modifications, it can be applied to practically all commercial metals.

Soft soldering

The mechanical strength of soft soldered sheet metal joints is usually in the order of 15–30 MN/m^2, and depends largely upon the nature of the solder used; the temperature at which the soldering is done; the depth of penetration of the solder, which in turn depends on capillary attraction, i.e., on the power of the heated surface to draw liquid metal through itself; the proper use of correctly designed soldering tools; the use of suitable fluxes; the speed of soldering and especially, workmanship.

Solders

Soft solder is an alloy of lead and tin and is used with the aid of a soldering flux. It is made from two base metals; tin and lead. Tin has a melting point of 232°C and lead of 327°C, but the alloy has a lower melting point than either of the two base metals and its lowest melting point is 183°C; this melting point may be raised by varying the percentage of lead or tin in the alloy. A small quantity of antimony is sometimes used in soft solder to increase its tenacity and improving its appearance by brightening the colour. The small percentage of antimony both improves the chemical properties of the solder and increases its tensile strength, without appreciably affecting its melting point or working properties.

Tech note

There is a great variety of solders, e.g., aluminium, bismuth, cadmium, silver, gold, pewterer's, plumber's, tinman's; solders are usually named according to the purpose for which they are intended.

The following solders are the most popular in use today:

95–100% *tin solder,* is used for high quality electrical work, where maximum electrical conductivity is required since the conductivity of pure tin is 20–40% higher than that of the most commonly used solders.

60/39.5/0.5 *(tin/lead/antimony) solder,* the eutectic composition, has the lowest melting point of all tin-lead solders and is quick setting. It also has the maximum bulk strength of all tin-lead solders and is used for fine electrical and tin smith's work.

50/47.5/2.5 *(tin/lead/antimony) solder,* called tin man's fine, contains more lead and is, therefore, cheaper than the 60/40 grade. Its properties in terms of low melting range and quick setting are still adequate, and hence it is used for general applications.

45/52.5/2.5 *(tin/lead/antimony) solder,* known as tin man's soft, is cheaper because of the higher lead content but has poorer wetting and mechanical properties. This solder is widely used for can soldering, where the maximum economy is required, and for any material which has already been tin plated, so that the inferior wetting properties of the solder are not critical.

30/68.5/1.5 *(tin/lead/antimony) solder,* known as plumber's solder, is also extensively used by the car body repairer. Because the material has a wide liquidus–solidus range (about 80°C),it remains in a pasty form for an appreciable time during cooling, and while in this condition it can be shaped or 'wiped' to form a lead pipe joint, or to the shape required for filling dents in frame tubes. Because of its high lead content, its wetting properties are very inferior and the surfaces usually have to be tinned with a solder of higher tin content first.

Fluxes

The function of flux is to remove oxides and tarnish from the metal to be joined so that the solder will flow, penetrate and bond to the metal surface; forming a good strong soldered joint. The hotter the metal, the more rapidly the oxide film forms. Without the chemical action of the flux on the metal, the solder would not tin the surface, and the joint would be weak and unreliable. Besides cleaning the metal, flux also ensures that no further oxidation from the atmosphere which could be harmful to the joint taking place during soldering, as this would restrict the flow of soldering.

Generally, soft soldering fluxes are divided into two main classes: corrosive fluxes and non-corrosive fluxes.

Tech note

Some fizzy drinks contain phosphoric acid—if you drop a dirty coin in the drink, the acid will clean it.

Brazing

Brazing is used extensively throughout the frame building trade as a quick and cheap means of joining frame tubes and other components. Although a brazed joint is not as strong as a fusion weld, it has many advantages which make it useful for the frame builder. Brazing is not classed as a fusion process and therefore, cannot be called welding, because the parent metals are not melted to form the joint but rely on a filler material of a different metal of low melting point which is drawn through the joint. The parent metals can be similar or dissimilar as long as the alloy rod has a lower melting point than either of them. The most commonly used alloy is of copper or zinc, which is, of course, brass. Brazing is accomplished by heating the pieces to be joined to a temperature higher than the melting point of the brazing alloy (brass). With the aid of flux, the melted alloy flows between the parts to be joined due to capillary attraction and diffuses into the surface of the metal, so that a strong joint is produced when the alloy cools. Brazing, or hard soldering to give it its proper name, is part fusion and is classed as a skin fusion process.

Tech note

Brazing is only possible on certain tube sets, check with your supplier before purchase.

Brazing is carried out at a much higher temperature than that required for the soft soldering process. A borax type of powder flux is used, which fuses to allow brazing to take place between 750 and 900°C. There are a wide variety of alloys in use as brazing rods; the most popular compositions contain copper in the ranges 46–50 and 58.5–61.5%, the remaining percentage being zinc.

The brazing process comprises the following steps:

1. Thoroughly clean the metal to be joined.
2. Using a welding torch, heat the metals to a temperature below their own critical or melting temperature. In the case of steel, the metal is heated to a dull cherry red.
3. Apply borax flux either to the rod or to the work as the brazing proceeds to reduce oxidation and to float the oxides to the surface.
4. Use the oxy-acetylene torch with a neutral flame, as this will give good results under normal conditions.

SAFETY NOTE

When heated zinc-plated steel (galvanised) gives off very toxic fumes, full respiratory equipment must be used — it is better to avoid this hazard if possible.

The main advantages of brazing are:

1. The relatively low temperature (750–900°C) necessary for a success-ful brazing job reduces the risk of distortion.
2. The joint can be made quickly and neatly, requiring very little clean-ing up.
3. Brazing makes possible the joining of two dissimilar metals; for example, brass can be joined to steel.
4. It can be used to repair parts that have to be re-chromed. For instance, a chromed fork which has been deeply scratched can be readily filled with brazing and then filed up ready for chroming.
5. Brazing is very useful for joining steels which have high carbon con-tent, or broken castings where the correct filler rod is not available.

Silver soldering

Silver solder probably originated in the manufacture and repair of silverware and jewellery for the purpose of securing adequate strength and the desired colour of the joint, but the technique of joining sheet metal products and components with silver solder was used for a long time on high-quality bicycle frames. The term 'soft soldering' has been widely adopted when referring to the older process to avoid confusion with the newer hard soldering process, known generally as either silver soldering or silver brazing. The use of silver solder on metals and alloys other than silver has grown largely because of the perfection by manufacturers of these solders which makes them easily appli-cable to many metals and alloys by means of the oxy-acetylene welding torch.

Solders and fluxes

Silver solders are more malleable and ductile than brazing rods, and hence joints made with silver solder have a greater resistance to bending stresses, shocks and vibration than those made with ordinary brazing alloys, as you can see this is very appropriate for bicycle frames. Silver solders are made in the strip, wire (rod) or granular form and several different grades of fus-ibility, the melting points vary between 630 and 800°C according to the percentages of silver, copper, zinc and cadmium they contain.

As in all non-fusion processes, the important factor is that the joint to be soldered must be perfectly clean. Hence, special care must be taken in preparing the metal surfaces to be joined with silver solder. Although fluxes will dissolve films of oxide during the soldering operation, frame tubes and lugs that are clean are much more likely to make a stronger, sounder joint than when impurities are present. The joints should fit closely and the parts must be held together firmly while being silver soldered because silver solders in the molten state are remarkably fluid and can penetrate minute spaces between the metals to be joined. The use of a frame building jig is essential with this process. In order to protect the metal surface against oxidation and to increase the flowing properties of the solder, a suitable flux such as borax or boric acid is used.

Silver soldering process

In silver soldering, the size of the welding tip used and the adjustment of the flame are very important to avoid overheating, as prolonged heating promotes oxide films; which weaken both the base metal and the joint material. This should be guarded against by keeping the luminous cone of the flame well back from the point being heated. When the joint has been heated just above the temperature at which the silver solder flows, the flame should be moved away and the solder applied to the joint, usually in rod form. The flame should then be played over the joint so that the solder and flux flow freely through the joint by capillary attraction. The finished silver soldered joint should be smooth, regular in shape and require no dressing up apart from the removal of the flux by washing in water.

When making a silver solder joint between dissimilar metals, concentrate the application of heat on the metal which has the higher heat capacity. This depends on the thickness and the thermal conductivity of the metals. The aim is to heat both members of the joint evenly, so that they reach the soldering temperature at the same time.

The most important points during the silver soldering are:

1. Cleanness of the joint surfaces.
2. Use of the correct flux.
3. The avoidance of overheating.

ALUMINIUM BRAZING

There is a distinction between the brazing of aluminium and the brazing of other metals. For aluminium, the brazing alloy is one of the aluminium alloys having a melting point below that of the parent metal. For other metals, the brazing alloys are often based on copper–zinc alloys (brasses — hence the term brazing) and are necessarily dissimilar in composition to the parent metal.

Wetting and fluxing

When a surface is wetted by a liquid, a continuous film of the liquid remains on the surface after draining. This condition, essential for brazing, arises when there is a mutual attraction between the liquid flux and solid metal due to a form of chemical affinity. Having accomplished its primary duty of removing the oxide film, the cleansing action of the flux restores the free affinities at the surface of the joint faces, promoting wetting by reducing the contact angle developed between the molten brazing alloy and parent metal. This action assists spreading and the feeding of brazing alloy to the capillary spaces, leading to the production of well-filled joints. An important feature of the brazing process is that the brazing alloy is drawn into the joint area by capillary attraction: the smaller the gap is between the two metal faces to be joined, the deeper is the capillary penetration.

The various grades of pure aluminium and certain alloys are amenable to brazing. Aluminium–magnesium alloys containing more than 2% magnesium are difficult to braze, as the oxide film is tenacious and hard to remove with ordinary brazing fluxes. Other alloys cannot be brazed because they start to melt at temperatures below that of any available brazing alloy. Aluminium–silicon alloys of nominal 5%, 7.5% or 10% silicon content are used for brazing aluminium and the alloy of aluminium and 1.5% manganese.

The properties required for an effective flux for brazing aluminium and its alloys are as follows:

1. The flux must remove the oxide coating present on the surfaces to be joined. It is always important that the flux be suitable for the parent metal, but especially so in the joining of aluminium–magnesium alloys.
2. It must thoroughly wet the surfaces to be joined so that the filler metal may spread evenly and continuously.
3. It must flow freely at a temperature just below the melting point of the filler metal.
4. Its density, when molten, must be lower than that of the brazing alloy.
5. It must not attack the parent surfaces dangerously in the time between its application and removal.
6. It must be easy to remove from the brazed assembly.

Many types of proprietary fluxes are available for brazing aluminium. These are generally of the alkali halide type, which are basically the mixtures of the alkali metal chlorides and fluorides. Fluxes and their residues are highly corrosive and therefore, must be completely removed after brazing by washing with hot water.

Brazing method

When the cleaned parts have been assembled, brazing flux is applied evenly over the joint surface of both parts to be brazed and the filler rod (brazing alloy). The flame is then played uniformly over the joint until the flux has dried and become first powdery, then molten and transparent. (At the powdery stage care is needed to avoid dislodging the flux, and it is often preferable to apply flux with the filler rod). When the flux is molten, the brazing alloy is applied, preferably from above, so that gravity assists in the flow of metal. In good practice, the brazing alloy is melted by the heat of the assembly rather than directly by the torch flame. Periodically, the filler rod is lifted and the flame is used to sweep the liquid metal along the joint; but if the metal is run too quickly in this way it may begin to solidify before it properly diffuses into the mating surfaces. The trial will show whether more than one feed point for the brazing alloy is necessary, but with proper fluxing, giving an unbroken path of flux over the whole joint width, a single feed is usually sufficient.

Bronze welding

Bronze welding is carried out much as in fusion welding except that the base metal is not melted. The base metal is simply brought up to tinning temperature (dull red colour) and a bead is deposited over the seam with a bronze filler rod. Although the base metal is never actually melted, the unique characteristics of the bond formed by the bronze rod are such that the results are often comparable with those secured through fusion welding. Bronze welding resembles brazing, but only up to a point. The application of brazing is generally limited to joints where a close fit or mechanical fastening serves to consolidate the assembly and the joint is merely strengthened or protected by the brazing material. In bronze welding, the filler metal alone provides the joint strength, and it is applied by the manipulation of a heating flame in the same manner as in gas fusion welding. The heating flame is made to serve the dual purpose of melting off the bronze rod and simultaneously heating the surface to be joined. The operator in this manner controls the work; hence the term 'bronze welding'.

Almost any copper–zinc alloy or copper–tin alloy or copper–phosphorus alloy can be used as a medium for such welding, but the consideration of costs, flowing qualities, strength and ductility of the deposit have led to the adoption of one general-purpose 60–40 copper–zinc alloy with minor constituents incorporated to prevent zinc oxide forming and to improve fluidity and strength. Silicon is the most important of these minor constituents, and its usefulness is apparent in three directions. First, in the manner with which it readily unites with oxygen to form silica, silicon provides a covering for the molten metal which prevents zinc volatilization and serves to maintain the balance of the constituents of the alloy; this permits the original high strength of the alloy to be carried through to the deposit. Second, this coating of silica combines with the flux used in bronze welding to form a very fusible slag, and this materially assists the tinning operation, which is an essential feature of any bronze welding process. Third, by its capacity for retaining gases in solution during solidification of the alloy, silicon prevents the formation of gas holes and porosity in the deposited metal, which would naturally reflect unfavourably upon its strength as a weld.

It is essential to use an efficient and correct flux. The objects of a flux are: first, to remove oxide from the edges of the metal, giving a chemically clean surface on to which the bronze will flow and to protect the heated edges from the oxygen in the atmosphere; second, to float oxide and impurities introduced into the molten pool to the surface, where they can do no harm. Although general-purpose fluxes are available, it is always desirable to use the fluxes recommended by the manufacturer of the particular rod being employed.

Bronze welding procedure

1. An essential factor for bronze welding is a clean metal surface. If the bronze is to provide a strong bond, it must flow smoothly and evenly over the entire weld area. Clean the surfaces thoroughly with a stiff wire brush. Remove all scale, dirt or grease; otherwise, the bronze will not adhere. If a surface has oil or grease on it, remove these substances by heating the area to a bright red colour and thus, burning the moff.
2. On thick sections, especially in repairing castings, bevel the edges to form a 90° V-groove. This can be done by chipping, machining, filing or grinding.
3. Adjust the torch to obtain a slightly oxidizing flame. Then heat the surfaces of the weld area.
4. Heat the bronzing rod and dip it in the flux. (This step is not necessary if the rods have been prefluxed.) In heating the rod, do not apply the inner cone of the flame directly to the rod.
5. Concentrate the flame on the starting end until the metal begins to turn red. Melt a little bronze rod on to the surface and allow it to spread along the entire seam. The flow of this thin film of bronze is known as the tinning operation. Unless the surfaces are tinned properly, the bronzing procedure to follow cannot be carried out successfully. If the base metal is too hot, the bronze will tend to bubble or run around like drops of water on a warm stove. If the bronze forms into balls which tend to roll off, just as water would if placed on a greasy surface, then the base metal is not hot enough. When the metal is at the proper temperature, the bronze spreads out evenly over the metal.
6. Once the base metal is tinned sufficiently, start depositing the proper size beads over the seam. Use a slightly circular torch motion and run the beads as in regular fusion welding with a filler rod. Keep dipping the rod in the flux as the weld progresses forward. Be sure that the base metal is never permitted to get too hot.
7. If the pieces to be welded are grooved, use several passes to fill the V. On the first pass, make certain that the tinning action takes place along the entire bottom surface of the V and about half-way upon each side. The number of passes to be made will depend on the depth of the V. When depositing several layers of beads, be sure that each layer is fused into the previous one.

HEALTH, SAFETY AND THE ENVIRONMENT

The materials used in soldering can be very dangerous if not used in a safe way. Besides the normal workshop precautions, the following should be especially noted:

- All soldering and brazing must be carried out in an area with suitable fume extraction.

- Fluxes are mainly corrosive and should be handled accordingly.
- Always wash your hands after soldering or brazing.
- Use PPE as directed by your company safety guidelines.
- All materials must be stored securely and not be accessible to unauthorised personnel.

SAFETY NOTE

You must remember that all gases are dangerous if not handled and stored correctly — always follow the manufacturer's instructions.

The following is a summary of gas characteristics and cylinder colour codes.

Oxygen

Cylinder colour: Black.
Characteristics: No smell. Generally considered non-toxic at atmospheric pressure. Will not burn but supports and accelerates combustion. Materials not normally considered combustible may be ignited by sparks in oxygen-rich atmospheres.

Nitrogen

Cylinder colour: Grey with a black shoulder.
Characteristics: No smell. Does not burn. Inert and hence, will cause asphyxiation in high concentrations.

Argon

Cylinder colour: Blue.
Characteristics: No smell. Heavier than air. Does not burn. Inert. Will cause asphyxiation in the absence of sufficient oxygen to support life. Will readily collect in the bottom of a confined area. At high concentrations, almost instant unconsciousness may occur, followed by death. The prime danger is that there will be no warning sign before unconsciousness occurs.

Propane

Cylinder colour: Bright red and bearing the words 'propane' and 'highly flammable'.
Characteristics: Distinctive fish-like offensive smell. Will ignite and burn instantly from a spark or piece of hot metal. It is heavier than air and will collect in ducts, drains or confined areas. Fire and explosion hazard.

Acetylene

Cylinder colour: Maroon.

Characteristics: Distinctive garlic smell. Fire and explosion hazard. Will ignite and burn instantly from a spark or piece of hot metal. It is lighter than air and less likely than propane to collect in confined areas. Requires minimum energy to ignite in air or oxygen. Never use copper or alloys containing more than 70% copper or 43% silver with acetylene.

Hydrogen

Cylinder colour: Bright red.

Characteristics: No smell. Non-toxic. Much lighter than air. Will collect at the highest point in any enclosed space unless ventilated there. Fire and explosion hazard. Very low ignition energy.

Carbon dioxide

Cylinder colour: Black, or black with two vertical white lines for liquid withdrawal.

Characteristics: No smell but can cause the nose to sting. Harmful. Will cause asphyxiation. Much heavier than air. Will collect in confined areas.

Argoshield

Cylinder colour: Blue with a green central band and a green shoulder.

Characteristics: No smell. Heavier than air. Does not burn. Will cause asphyxiation in the absence of sufficient oxygen to support life. Will readily collect at the bottom of confined areas.

SAFETY MEASURES

General gas storage procedures are listed below:

1. Any person in charge of the storage of compressed gas cylinders should know the regulations covering highly flammable liquids and compressed gas cylinders as well as the characteristics and hazards associated with individual gases.
2. It is best to store full or empty compressed gas cylinders in the open, in a securely fenced compound, but with some weather protection.
3. Within the storage area, oxygen should be stored at least 3 m from fuel gas supply.
4. Full cylinders should be stored separately from the empties and cylinders of different gases, whether full or empty, should be segregated from each other.
5. Other products must not be stored in a gas store, particularly oils or corrosive liquids.
6. It is best to store all cylinders upright, taking steps, particularly with round-bottomed cylinders, to see that they are secured to prevent

them from falling. Acetylene and propane must *never* be stacked horizontally in storage or in use.

7. Storage arrangements should ensure adequate rotation of stock.

Acetylene cylinders

1. The gas is stored together with a solvent (acetone) in maroon painted cylinders, at a pressure of 17.7 bar maximum at 15°C. The cylinder valve outlet is screwed left-handed.
2. The hourly rate of withdrawal from the cylinder must not exceed 20% of its content.
3. Pressure gauges should be calibrated up to 40.0 bar.
4. As the gas is highly flammable, all joints must be checked for leaks using soapy water.
5. Acetylene cylinders must be stored and used in an upright position and protected from excessive heat and coldness.
6. Acetylene can form explosive compounds in contact with certain metals and alloys, especially those of copper and silver. Joint fittings made of copper should not be used under any circumstances.
7. The colour of cylinders, valve threads, or markings must not be altered or tampered within any way.

Oxygen cylinders

1. This gas is stored in black painted cylinders at a pressure of 200/230 bar, maximum at 15°C.
2. Never under any circumstances allow oil or greases to come into contact with oxygen fittings because spontaneous ignition may take place.
3. Oxygen must not be used in place of compressed air.
4. Oxygen escaping from a leaking hose will form an explosive mixture with oil or grease.
5. Do not allow cylinders to come into contact with electricity.
6. Do not use cylinders as rollers or supports.
7. Cylinders must not be handled roughly, knocked or allowed to fall to the ground.

GENERAL EQUIPMENT SAFETY

All equipment should be subjected to regular periodic examination and overhaul. Failure to do so may allow equipment to be used in a faulty state and may be dangerous.

Rubber hose: Use only hoses in good condition, fitted with the special hose connections attached by permanent ferrules. Do not expose the hose to heat, traffic, slag, sparks from welding operations, or oil or grease. Renew the hose as soon as it shows any sign of damage.

Pressure regulators: Always treat a regulator carefully. Do not expose it to knocks, jars or sudden pressure caused by the rapid opening of the cylinder valve. When shutting down, release the pressure on the control spring after the pressure in the hoses has been released. Never use a regulator on any gas except that for which it was designed, or for higher working pressures. Do not use regulators with broken gauges.

Welding torch: When lighting up and extinguishing the welding torch, the manufacturer's instructions should always be followed. To clean the nozzle. use special nozzle cleaners; never a steel wire.

Fluxes: Always use welding fluxes in a well-ventilated area.

Goggles: These should be worn at all times during welding, cutting or merely observing.

Protection: Leather or fire-resistant clothing should be worn for all heavy welding or cutting. The feet should be protected from sparks, slag or cut material falling on them.

GAS SHIELDED ARC WELDING (MIG, MAG AND TIG)

Development of gas shielded arc welding

Originally, the process was evolved in America in 1940 for welding in the aircraft industry. It developed into the TIG shielded arc process, which in turn, led to shielded MIG arc welding. The latter became established in this country in 1952.

In the gas shielded arc process, heat is produced by the fusion of an electric arc maintained between the end of a metal electrode, either consumable or non-consumable, and the part to be welded, with a shield of protective gas surrounding the arc and the weld region. There are at present in use three different types of gas shielded arc welding:

TIG: The arc is struck by a non-consumable tungsten electrode and the metal to be welded, and filler metal is added by feeding a rod by hand into the molten pool.

MIG: This process employs a continuous feed electrode which is melted in the intense arc heat and deposited as weld metal: hence, the term consumable electrode. This process uses only inert gases, such as argon and helium, to create the shielding around the arc.

MAG: This is the same as MIG except that the gases have an active effect upon the arc and are not simply an inert envelope. The gases used are carbon dioxide or argon/carbon-dioxide mixtures.

Tech note

Gas tungsten arc welding is the terminology used in America and many parts of Europe for the TIG welding process, and it is becoming increasingly accepted as the standard terminology for this process.

Chapter 11

Reinforced composite materials

INTRODUCTION

Composite materials came into use in the automotive and boat building industries in the 1950s to fulfil the needs of small post-war car manufacturers, as metal was rationed for 12 years after the end of World War II small companies sought alternatives. The automotive market was developing at the same time as the small boat market; both developed along the same lines using a glass fibre and a resin lay-up procedure called glass reinforced plastics (GRP), which is still used today by many kit car manufacturers and makers of human powered vehicles (HPVs). The use of **composites—a product made up of more than one material which is bonded together to provide special properties**—became more specialized as **carbon-based technical materials** became available. Carbon fibre, as it is called, was originally invented and patented by The Royal Aircraft Establishment at Farnborough, UK. Bicycles started to be made from it in the 1990s as the materials became more readily available and methods of joining tubes and laying up became known. Given the use of moulds and autoclave equipment, the manufacture of frames in carbon fibre should be cheaper and require less skill than the construction of frames from metal.

Depending on the materials used in composite construction, the following properties of composites may influence their choice:

- Components can be produced on a one-off basis with minimum tooling.
- Compound curvature can be produced with constant material thickness.
- Extreme lightness for a given strength.
- Resistant to corrosion.
- Different finishes are available.
- There is a style cache in the use of carbon fibre.

BASIC PRINCIPLES OF REINFORCED
COMPOSITE MATERIALS

The basic principle involved in reinforced plastic production is the combination of polyester resin and reinforcing fibres to form a solid structure. GRP are essentially a family of structural materials, which utilize a very wide range of thermoplastic and thermosetting resins. The incorporation of glass fibres in the resins changes them from relatively low strength, brittle materials into strong and resilient structural materials. In many ways, glass fibre reinforced plastic can be compared to concrete, with the glass fibres performing the same function as the steel reinforcement and the resin matrix acting as the concrete. Glass fibres have high strength and high modulus, and the resin has low strength and low modulus. Despite this, the resin has the important task of transferring the stress from fibre to fibre, so enabling the glass fibre to develop its full strength.

Polyester resins are supplied as viscous liquids which solidify when the actuating agents in the form of a catalyst and accelerator are added. The proportions of this mixture, together with the existing workshop conditions, dictate whether it is cured at room temperature or at higher temperatures and also the length of time needed for curing. In common practice, pre-accelerated resins are used, requiring only the addition of a catalyst to affect the cure at room temperature. Glass reinforcements are supplied in a number of forms, including chopped strand mats, needled mats, bidirectional materials such as woven rovings and glass fabrics and rovings which are used for chopping into random lengths or as high-strength directional reinforcement. Other materials needed are the releasing agent, filler and pigment concentrates for the colouring of glass fibre reinforced plastic.

Among the methods of production, the most used method is that of contact moulding, or the wet laying-up technique as it is sometimes called. The mould itself can be made of any material which will remain rigid during the application of the resin and glass fibre, which will not be attacked by the chemicals involved and which will also allow easy removal after the resin has set hard. Those in common use are wood, plaster, sheet metal and glass fibre itself, or a combination of these materials. The quality of the surface of the completed moulding will depend entirely upon the surface finish of the mould from which it is made. When the mould is ready, the releasing agent is applied followed by a thin coat of resin to form a gel coat. To this, a fine surfacing tissue of fibre glass is often applied. Further, resin is applied, usually by brush, and carefully cut-out pieces of mat or woven cloth are laid in position. The use of split washer rollers removes the air and compresses the glass fibres into the resin. Layers of resin and glass fibres are added until the required thickness is achieved. Curing takes place at room temperature, but heat can be applied to speed up the curing time. Once the catalyst has caused the resin to set hard, the moulding can be taken from the mould.

MANUFACTURE OF REINFORCED COMPOSITE MATERIALS

When glass is drawn into fine filaments, its strength greatly increases over that of bulk glass. Glass fibre is one of the strongest of all materials. The ultimate tensile strength of a single glass filament (diameter 9–15 micrometres) is about 3447000 kN/m^2. It is made from readily available raw materials, and is non-combustible and chemically resistant. Glass fibre is therefore, the ideal reinforcing material for plastics. In Great Britain, the type of glass which is principally used for glass fibre manufacture is E glass, which contains less than1% alkali borosilicate glass. E glass is essential for electrical applications and it is desirable to use this material, where good weathering and water resistance properties are required. Therefore, it is greatly used in the manufacture of composite HPV body shells.

Basically the glass is manufactured from sand or silica and the process by which it is made proceeds through the following stages:

1. Initially the raw materials, including sand, china clay and limestone, are mixed together as powders in the desired proportions.
2. The 'glasspowder', or frit as it is termed, is then fed into a continuous melt furnace or tank.
3. The molten glass flows out of the furnace through a forehearth to a series of fiberizing units usually referred to as bushings, each containing several hundreds of fine holes. As the glass flows out of the bushings under gravity, it is attenuated at high speed.

After fiberizing, the filaments are coated with a chemical treatment usually referred to as a forming size. The filaments are then drawn together to form a strand, which is wound on a removable sleeve on a high-speed winding head. The basic packages are usually referred to as cakes and form the basic glass fibre which, after drying, is processed into the various reinforcement products. Most reinforcement materials are manufactured from continuous filaments ranging in fibre diameter from 5 to 13 micrometres. The fibres are made into strands by the use of size. In the case of strands, which are subsequently twisted into weaving yarns, the size lubricates the filaments as well as acting as an adhesive. These textile sizes are generally removed by heat or solvents and replaced by a chemical finish before being used with polyester resins. For strands which are not processed into yarns, it is usual to apply sizes which are compatible with moulding resins.

Glass reinforcements are supplied in a number of forms, including chopped strand mats (CSMs), needled mats, bi-directional materials such as woven rovings and glass fabrics and rovings, which are used for chopping into random lengths or as high-strength directional reinforcements.

TYPES OF REINFORCING MATERIAL

Woven fabrics

Glass fibre fabrics are available in a wide range of weaves and weights. Light weight fabrics produce laminates with higher tensile strength and modulus than heavy fabrics of a similar weave. The type of weave will also influence the strength (due, in part, to the amount of crimp in the fabric), and usually satin weave fabrics, which have little crimp, give stronger laminates than plain weaves which have a higher crimp. Satin weaves also drape more easily and are quicker to impregnate. Besides fabrics made from twisted yarns, it is now the practice to use woven fabrics manufactured from rovings. These fabrics are cheaper to produce and can be much heavier in weight.

Chopped strand mat (CSM)

CSM is the most widely used form of reinforcement. It is suitable for moulding the most complex forms. The strength of laminates made from chopped strand mat is less than that with woven fabrics, since the glass content which can be achieved is considerably lower. The laminates have similar strengths in all directions because the fibres are random in orientation. Chopped strand mat consists of randomly distributed strands of glass about 50 mm long which are bonded together with a variety of adhesives. The type of binder or adhesive will produce differing moulding characteristics and will tend to make one mat more suitable than another for specific applications.

Needle mat

This is mechanically bound together and the need for an adhesive binder is eliminated. This mat has a high resin pick-up owing to its bulk, and cannot be used satisfactorily in moulding methods, where no pressure is applied. It is used for press moulding and various low-pressure techniques such as pressure injection, vacuum and pressure bag.

Rovings

These are formed by grouping untwisted strands together and winding them on a 'cheese'. They are used for chopping applications to replace mats either in contact moulding (spray-up), or translucent sheet manufacture of press moulding (pre-form). Special grades of roving are available for each of these different chopping applications. Rovings are also used for weaving, for filament winding and for pultrusion processes. Special forms are available to suit these processes.

Chopped strands

These consist of rovings pre-chopped into strands of 6 mm, 13 mm, 25 mm or 50 mm lengths. This material is used for dough moulding compounds, and in casting resins to prevent cracking.

Staple fibres

These are occasionally used to improve the finish of mouldings. Two types are normally available, a compact form for contact moulding and a soft bulky form for press moulding. These materials are frequently used to reinforce gel coats. The weathering properties of translucent sheeting are considerably improved by the use of surfacing tissue.

Application of these materials

Probably chopped strand mat is the most commonly used for the average moulding. It is available in several different thicknesses and specified by weight: 300, 450 and 600 g/m^2. The 450 g/m^2 is the most frequently used, and is often supplemented with the 300 g/m^2. The 600 g/m^2 density is rather too bulky for many purposes, and may not drape as easily, although all forms become very pliable, when wetted with the resin.

The woven glass fibre cloths are generally of two kinds, made from continuous filaments or from staple fibres. Obviously, most fabricators use the woven variety of glass fibre for those structures that are going to be the most highly stressed. For example, a moulded glass fibre seat pan and squab unit in a HPV would be made with woven material as reinforcement, but a detachable hard top for a trailer body would more probably be made with chopped strand mat as a basis. However, it is quite customary to combine cloth and mat, not only to obtain adequate thickness, but because if the sandwich of resin, mat and cloth is arranged, so that the mat is the nearest to the surface of the final product, the appearance will be better.

The top layer of resin is comparatively thin, and the weave of cloth can show up underneath it, especially if some areas have to be buffed subsequently. Chopped fibres do not show up so prominently, but some fabricators compromise by using the thinnest possible cloth (surfacing tissue as it is known) nearest the surface, on top of the chopped strand mat. When moulding, these orders are of course reversed, the tissue going onto the gel coat on the inside of the mould, followed by the mat and resin lay-up.

It is important to note that if glass cloths or woven mat are used, it is possible to lay up the materials so that the reinforcement is in the direction of the greatest stresses, thus giving extra strength to the entire fabrication. In plain weave cloths, each warp and weft thread passes over one yarn and under the next. In twill weaves, the weft yarns pass over one warp and under more than one warp yarn; in 2 $\times$ 1 twill, the weft yarns pass over one warp yarn and under two warp yarns. Satin weaves may be of multishaft types, when each warp and weft yarn goes under one and over several yarns. Unidirectional cloth is one in which the strength is higher in one direction than the other, and balanced cloth is a type with the warp and weft strength about equal. Although relatively expensive, the woven forms have many excellent qualities, including high dimensional stability, high tensile and impact strength, good heat, weather and chemical resistance, good moisture absorption, resistance to fire and good thermo-electrical properties. A number of different weaves and weights are available, and thickness may range from 0.05 mm to 9.14 mm, with weights from 30 g/m^2 to 1 kg/m^2, although the grades mostly used in the automotive field probably have weights of about 60 g/m^2 and will be of plain, twill or satin weave.

Carbon fibre

This is another reinforcing material. Carbon fibres possess a very high modulus of elasticity, and have been used successfully in conjunction with epoxy resin to produce low-density composites possessing high strength.

RESINS USED IN REINFORCED COMPOSITE MATERIALS

The first man-made plastics were produced in this country in 1862 by Alexander Parkes and were the fore runner of celluloid. Since then, a large variety of plastics have been developed commercially, particularly in the last 25 years. They extend over a wide range of properties. Phenol formaldehyde is a hard thermoset material; polystyrene is a hard, brittle thermoplastic; polythene and plasticized polyvinyl chloride (PVC) are soft, tough thermoplastic materials and so on. Plastics also exist in various physical forms. They can be bulk solid materials, rigid or flexible foams or in the form of sheet or film. All plastics have one important common property. They are composed of macro-molecules, which are large chain-like molecules consisting of many simple repeating units. The chemist calls these molecular chains polymers. Not all polymers are used for making plastic mouldings. Man-made polymers are called synthetic resins until they have been moulded in some way, when they are called plastics.

Most synthetic resins are made from oil. The resin is an essential component of glass fibre reinforced plastic. The most widely used is unsaturated polyester resin, which can be cured to a solid state either by catalyst and heat or by catalyst and accelerators at room temperature. The ability of polyester resin to cure at room temperature into a hard material is one of the main reasons for the growth of the reinforced plastics industry. It was this which led to the development of the room temperature contact moulding methods, which permit production of extremely large integral units.

Tech note

Scientists are working on making carbon fibre materials bio-degradable.

Polyester resins are formulated by the reaction of organic acids and alcohols which produce a class of materials called esters. When the acids are polybasic and the alcohols are polyhydric, they can react to form a very complex ester, which is generally known as polyester. These are usually called alkyds, and have long been important in surface coating formulations because of their toughness, chemical resistance and endurance. If the acid or alcohol used contains an unsaturated carbon bond, the polyester formed can react further with other unsaturated materials such as styrene or diallyl phthalate. The result of this reaction is to interconnect the different polyester units to form the three-dimensional cross-linked structure that is characteristic of thermosetting resins. The available polyesters are solutions of these alkyds in the cross-linking monomers. The curing of the resin is the reaction of the monomer and the alkyd to form the cross-linked structure. An unsaturated polyester resin is one which is capable of being cured from a liquid to a solid state when subjected to the right conditions. It is usually referred to as polyester.

Catalysts and accelerators

In order to mould or laminate a polyester resin, the resin must be cured. This is the name given to the overall process of gelation and hardening, which is achieved either by the use of a catalyst and heating, or at normal room temperature by using a catalyst and an accelerator. Catalysts for polyester resins are usually organic peroxides. Pure catalysts are chemically unstable and liable to decompose with explosive violence. They are supplied, therefore, as a paste or liquid dispersion in a plasticizer, or as a powder in inert filler. Many chemical compounds act as accelerators, making it possible for the resin-containing catalyst to be cured without the use of heat. Some accelerators have only limited or specific uses, such as quaternary ammonium compounds, vanadium, tin or zirconium salts. By far the most important of all accelerators are those based on cobalt soap or those based on a tertiary amine. It is essential to choose the correct type of catalyst and accelerator as well as to use the correct amount, if the optimum properties of the final cured resin or laminate are to be obtained.

Pre-accelerated resins

Many resins are supplied with an in-built accelerator system controlled to give the most suitable gelling and hardening characteristics for the fabricator. Pre-accelerated resins need only the addition of a catalyst to start the curing reaction at room temperature. Resins of this type are ideal for production runs under controlled workshop conditions.

The cure of a polyester resin will begin as soon as a suitable catalyst is added. The speed of the reactions will depend on the resin and the activity of the catalyst. Without the addition of an accelerator, heat or ultraviolet radiation, the resin will take a considerable time to cure. In order to speed up this reaction at room temperature, it is usual to add an accelerator. The quantity of accelerator added will control the time of gelation and the rate of hardening.

There are three distinct phases in the curing reaction:

Gel time: This is the period from the addition of the accelerator to the setting of the resin to a soft gel.

Hardening time: This is the time from the setting of the resin to the point when the resin is hard enough to allow the moulding or laminate to be withdrawn from the mould.

Maturing time: This may be hours, several days or even weeks depending on the resin and curing system, and is the time taken for the moulding or laminate to acquire its full hardness and chemical resistance. The maturing process can be accelerated by post-curing. When the material is not fully matured, it is referred to as being in its **green state,** or simply as **green;** this term is taken from the colour of the wood of a freshly chopped down tree.

Fillers and pigments

Fillers are used in polyester resins to impart particular properties. They will give opacity to castings and laminates, produce dense gel coats and impart specific mechanical, electrical and fire resisting properties. A particular property may often be improved by the selection of suitable filler. Powdered mineral fillers usually increase compressive strength; fibrous fillers improve tensile and impact strength. Moulding properties can also be modified by the use of fillers; for example, shrinkage of the moulding during cure can be considerably reduced. There is no doubt, also, that the wet lay-up process on vertical surfaces would be virtually impossible, if thixotropic fillers were not available.

Polyester resins can be coloured to any shade by the addition of selected pigments and pigment pastes, the main requirement being to ensure thorough dispersion of colouring matter throughout the resin to avoid patchy mouldings.

Both pigments and fillers can increase the cure time of the resin by dilution effect, and the adjusted catalyst and promoter are added to compensate.

Releasing agents

Releasing agents used in the normal moulding processes may be either water-soluble film-forming compounds, or some type of wax compound.

The choice of releasing agent depends on the size and complexity of the moulding and on the surface finish of the mould. Small mouldings of simple shape, taken from a suitable GRP mould, should require only a film of polyvinyl alcohol (PVAL) to be applied as a solution by cloth, sponge or spray. Some mouldings are likely to stick if only PVAL is used. PVAL is available as a solution in water or solvent, or as a concentrate which has to be diluted, and it may be in either coloured or colourless form.

Suitable wax emulsions are also available as a releasing agent. They are supplied as surface finishing pastes, liquid wax or wax polishes. The recommended method of application can vary depending upon the material to be finished. Hand apply with a pad of damp, good quality mutton cloth or equivalent, in straight even strokes. Buff lightly to a shine with a clean, dry, good quality mutton cloth. Machine at 1800 rpm using a G-mop foam finishing head. Soak this head in clean water before use and keep damp during compounding, apply the wax to the surface. After compounding, remove residue and buff lightly to a shine with a clean, dry, good quality mutton cloth.

Wax polishes should be applied in small quantities, since they contain a high percentage of wax solids. Application with a pad of clean, soft cloth should be limited to an area of approximately 1 square metre. Polishing should be carried out immediately, before the wax is allowed to dry. This can be done either by hand or by machine with the aid of a wool mop polishing bonnet.

Frekote is a semi-permanent, multi-release, gloss finish, non-wax polymeric mould release system specially designed for high-gloss polyester mouldings. It will give a semi-permanent release interface, when correctly applied to moulds from ambient up to 135°C. This multi-release interface chemically bonds to the mould's surface and forms on it a micro-thin layer of a chemically resistant coating. It does not build up on the mould and will give a high-gloss finish to all polyester resins, cultured marble and onyx. It can be used on moulds made from polyester, epoxy, metal or composite moulds. Care should be taken to avoid contact with the skin, and the wearing of suitable clothing, especially gloves, is highly recommended. These products must be used in a well ventilated area.

Adhesives used with GRP

Since polyester resin is highly adhesive, it is the logical choice for bonding most materials to GRP surfaces.

Suitable alternatives include the Sika Technique, which is a heavy-duty, polyurethane-based joining compound. It cures to a flexible rubber which bonds firmly to wood, metal, glass and GRP. It is ideal for such jobs as bonding glass to GRP or bonding GRP and metal, as is often required on HPVs with GRP bodywork. It is not affected by vibration and is totally waterproof. The Araldite range includes a number of industrial adhesives which are highly recommended for use with GRP. Most high-strength impact adhesives (superglues) can be used on GRP laminates.

Most other adhesives will be incapable of bonding strongly to GRP and should not be used, when maximum adhesion is essential.

Core materials

Core materials, usually polyurethane, are used in sandwich construction that is basically a laminate consisting of a foam sheet between two or more glass fibre layers. This gives the laminate considerable added rigidity without greatly increasing weight. Foam materials are available which can be bent and folded to follow curved surfaces such as bicycle parts. Foam sheet can be glued or stapled together, then laminated over to produce a strong box structure, without requiring a mould. Typical formers and core materials are paper rope, polyurethane rigid foam sheet, scoreboard contoured foam sheet, Termanto PVC rigid foam sheet, Term PVC contoured foam sheet and Termino PVC contoured foam sheet.

Formers

A former is anything which provides shape or form to a GRP laminate. They are often used as a basis for stiffening ribs or box sections. A popular material for formers is a paper rope, made of paper wound on flexible wire cord. This is laid on the GRP surface and is laminated over to produce reinforcing ribs, which give added stiffness with little extra weight. The former itself provides none of the extra stiffness; this results entirely from the box section of the laminate rib. Wood, metal or plastic tubing and folded cardboard can all be used successfully as formers. Another popular material is polyurethane foam sheet, which can be cut and shaped to any required form.

Composite theory

In its most basic form, a composite material is one which is composed of two elements working together to produce material properties that are different to the properties of those elements on their own. In practice, most composites consist of a bulk material called the matrix, and a reinforcement material of some kind which increases the strength and stiffness of the matrix.

Polymer matrix composites are the type of composites used in modern vehicle bodywork. This type of composite is also known as fibre-reinforced polymers (or plastics). The matrix is a polymer-based resin and the reinforcement material is a fibrous material such as glass, carbon or aramid. Frequently, a combination of reinforcement materials is used.

The reinforcement materials have high tensile strength, but are easily chaffed and will break if folded. The polymer matrix holds the fibres in place, so that they are in their strongest position and protect them from damage.

The properties of the composites are thus determined by:

- The properties of the fibre.
- The properties of the resin.
- The ratio of fibre to resin in the composite—**fibre volume fraction (FVF)**.
- The geometry and orientation of the fibres in the composite.

Resin

The choice of resins depends on a number of characteristics, namely:

- **Adhesive properties:** In relation to the type of fibres being used and if metal inserts are to be used such as for panel fitting.
- **Mechanical properties:** Particularly tensile strength and stiffness.
- **Micro-cracking resistance:** Stress and age hardening cause the material to crack; the micro-cracks reduce the material strength and eventually lead to failure.
- **Fatigue resistance:** Composites tend to give better fatigue resistance than most metals.
- **Degradation from water ingress:** All laminates permit very low quantities of water to pass through in a vapour form. If the laminate is wet for a long period, the water solution inside the laminate will draw in more water through the osmosis process.
- **Curing properties:** The curing process alters the properties of the material. Generally, oven curing at between 80°C and 180°C will increase the tensile strength by up to 30%.
- **Cost:** The different materials cost different prices.

The main types of resins are: polyesters, vinylesters, epoxies, phenolics, cyanate esters, silicones, polyurethanes, bismaleides (**BMI**) and polyamides. The first three are the ones mainly used for manufacturing work as they are reasonably priced. Cyanates, BMI and polyamides cost about 10 times the price of the others.

Reinforcing fibres

The mechanical properties of the composite material are usually dominated by the contribution of the reinforcing fibres. The four main factors which govern this contribution are:

1. The basic mechanical properties of the fibre.
2. The surface interaction of the fibre and the resin—called the interface.
3. The amount of fibre in the composite—**FVF**.
4. The orientation of the fibres.

The three main reinforcing fibres used in HPVs are: glass, carbon and aramid. In addition, the following are used for non-body purposes: polyester, polyethylene, quartz, boron, ceramic and natural fibres such as jute and sisal.

Aramid fibre is a man-made organic polymer, an aromatic polyamide, produced by spinning fibre from a liquid chemical blend. The bright golden yellow fibres have high strength and low density giving a high specific strength. Aramid has good impact resistance. Aramid is better known by its Dupont trade name Kevlar.

Carbon fibre is produced by the controlled oxidation, carbonization and graphitization of carbon-rich organic materials—referred to as precursors—which are in fibre form. The most common precursor is polyacrylonitrile; pitch and cellulose are also used.

Pre-impregnated material (Pre-preg)

Woven material is available pre-impregnated with resin. It is referred to as Pre-preg. This means that the material has exactly the right amount of resin applied to it. The resin is fully coating the material—so that there are no dry spots, which could lead to component failure. Pre-preg is, therefore, quicker to use and the resin density is accurate.

Pre-preg has a limited shelf life which is compounded by the fact that it must be stored at −18°C. A deep freeze cabinet is, therefore, needed for storage. The Pre-preg cannot be unrolled nor cut when it is in the frozen state, so it must be removed from the freezer and brought up to normal room temperature. It is only possible to freeze and de-frost the Pre-preg a limited number of times, so the material must be managed carefully. The usual way to do this is by means of a control card. The dates and times of defrosting are recorded as is the amount of material taken off the roll. That way the life of the roll and the amount of material left can be seen without removing the roll from the freezer.

Curing

The resin, whether it is by wet lay-up or Pre-preg needs time and heat to dry it out and make it hard. When the hardener is added to the resin, it will generate heat chemically. Be careful, this heat can cause fire and other damage. However, at normal temperature, 20°C, it will take about 5 days for the resin to become fully hard. During this time period, the component should not be moved nor should any stress be applied. To speed up the hardening process and to add extra strength to the component, it is normal to use an oven. The oven may be a simple box with heating element, or an autoclave which is a cylindrical-shaped oven that can be pressurized or evacuated inside. The normal procedure is to place the newly made component in the oven, or autoclave, then rack up the temperature gently, over a period of about 30 minutes. Maintain the temperature typically at 150°C for about 5 hours and then gradually lower the temperature again over about a 30-minute period. The best way to do this is with a computer control system.

Core materials

Engineering theory tells us in most cases that the stiffness of a panel is proportional to the cube of its thickness, that is, the further apart that we can keep the outer fibres, the stiffer the panel will be. Putting a low density core between two layers of composite material will add stiffness with minimum weight and at reasonable cost.

Foam

A variety of materials are used, one of the most common is foam. Foam can be made from a variety of synthetic polymers. Densities of foam can vary between 30 and 300 kg/m^3 and thicknesses available are from 5 to 50 mm.

Honeycomb

Honeycombs are made from a variety of materials, including extruded thermoplastic—acrylonitrile butadiene styrene, polycarbonate, polypropylene and polyethylene—bonded paper, aluminium alloy and for fire-resistant parts, Nomex. Nomex is a paper-like material based on Kevlar fibres.

Heat

A point to be noted is that, the most carbon fibre materials are affected by heat. Thermal expansion can lead to micro-cracking. A carbon fibre panel, which is painted black, will absorb a lot of heat from the sun for a long period. This can cause the panel to expand, which could lead to micro-cracks in the panel and cracks in the paint work. This will then allow in moisture, which will cause further deterioration of the panel.

SAFETY NOTE

Because of the high level of stress applied to composite bicycle frames, attempts to repair are NOT RECOMMENDED.

Chapter 12

Data

Just about every individual and organization collect data of some sort or another. At a simple level, it may be checking the instruction sheet to see what the tyre pressures should be, or looking at the train time-table to find the next train to get to the office.

Tech note

Data, noun plural, facts and statistics collected together for reference or analysis. The singular is datum. Data includes statistics and information.

Bicycle riders collect the data for such things as mileage covered, or the best time for a 10-mile time trial on a favourite course. However, much more data is available for riders, trainers, company owners, sports scientist and anybody else interested bicycles. How you use it can give you a competitive edge, either in the sport or the industry.

THE NORM

The norm is something which is usual, typical or standard.

The secret in the use of data is actually knowing what it shows and what you can do with that information. In other words, what is the norm and what are the extremes, what makes them these cases and how can we utilize this information. We'll start with an explanation of the norm. If you are out with your cycling club, you will note that the frame sizes of your fellow riders vary, as the height and other statistics of the riders also vary. In fact, this concept of the norm came about through population studies—scientists visiting other countries many years ago and comparing the height and other attributes of the inhabitants of these countries. They would then say the norm for the height of people from country A is x so that they could compare them in country B, which is y. So, they developed the concept of the normal distribution curve. That is to say, not all people in country A are the same height, but the bulk of the population will vary within a few centimetres either side of x. Besides, there will be both very tall people and very short ones—those outside the norm. If you go into any high street clothes shop, you'll find that they usually only stock a limited range of sizes—the ones which fit those within the norm. Often this range is simply small, medium and large in their ratings and varies with manufactures.

UNDERSTANDING AND USING DATA

To use data, it is a good idea to consider Bloom's Taxonomy. It states the six levels of the cognitive domain. Most people work in the lower three levels, these are **Knowledge, Comprehension** and **Application** in Bloom's terminology. In other words, understanding the description of the data, being able to describe and discuss the actual content of the data and being able to apply the actual data in a real situation. For instance, considering your time trial results, over a year you will know on which days you performed best, what the weather was like or what course you were on.

The three higher levels are **Analysis, Synthesis and Evaluation.** Analysis means breaking the data down, understanding both the content and the structure. In terms of your time trial results, it is considering all the factors: bicycle used, gearing, clothing, course, weather, distance, time of day, other riders and what you had for breakfast amongst a myriad of other items. Synthesis is picking out the bits which you think were most important and re-mixing them to form a new structure. Evaluation is making a judgement on whether the new structure worked—then making more changes to improve it again. It is a continuous process.

RECORDING DATA

Recording data is very important, how you record it will control the way in which you can use it. Businesses such as takeaway shops use data to control their opening hours to match their customer needs. For instance, rarely will you find a chip-shop open before 12:00 pm, but London Kebab-shops often stay open till 4:00 am.

Don't waste time on recording data which you will not use, and remember that if it is not recent, it won't be relevant. Mood, fashions and tastes change quickly; as does technology and many other factors.

TALLY CHART

A simple way of collecting data is the use of a tally chart, the first column has the options listed. The second column is the tally, this is made in pencil line strikes up to four, then the fifth is across. The third column is the tally totalled.

If it is the number of attendees at an event a simple manual tally counter is ideal, push the lever for each head counted. The total number will be displayed.

SPREADSHEET

Spreadsheets are a great way of recording data, and of course, you can carry out detailed analysis, synthesis and evaluation using what-if scenarios on them. That is inserting projected values in the columns and seeing what happens to the results. This is of particular value when you are concerned with data changes of very small margins, say less than one percent. For example, changing your gearing by one tooth and retaining the same cadence rate, how much will this shave off your 10-mile time trial time. What the extra profit will be increasing the price of something by a few pence will make to the bottom line.

PRESENTING DATA

Spreadsheets are good for the recording and the analysis of data; but to get your message over, to both yourself and others, a visual form is useful. Print them out and pin them up on the wall to help motivate yourself and your staff or colleagues.

BAR CHARTS AND STACKING BAR CHARTS

The bar charts and stacking bar charts are good to show growth and changes in emphasis. The best approach is probably to make them in colour.

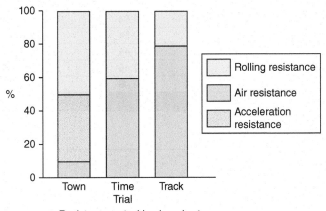

Figure 12.1 Stacking bar chart.

PIE CHARTS

The Pie chart is a really good way of showing percentage data as it gives a clear indication of percentages. A circle is made up of 360°, so 10% is 36°. To find, for example, 30% as an angle for the Pie chart:

$$30 / 100 \times 360° = 108°$$

NORMAL DISTRIBUTION

Calculations for normal distribution are simple to do, the following gives you two examples, ungrouped and grouped data use different methods. It is useful to be able to calculate the mean (average), the variance from the mean and the standard deviation.

UNGROUPED DATA

The following is a table of voltages measured across the lighting circuit of a low voltage system in a bicycle workshop to check volt-drop.

Sample	Voltages	Differences from mean	Differences squared
1	119	1.46	2.13
2	120	0.46	0.211
3	120	0.46	0.211
4	120	0.46	0.211
5	121	0.54	0.291
6	119	1.46	2.13
7	122	1.54	2.37
8	122	1.54	2.37
9	123	2.54	6.45
10	123	2.54	6.45
11	119	1.46	2.13
12	118	2.46	6.05
13	120	0.46	0.211
Total: 13	Total: 1566		Total: 31.215

Mean:

$$\bar{x} = \frac{Total\ of\ voltages}{Number\ in\ sample} = \frac{1566}{13} = 120.46$$

Variance:

$$S = \frac{Sum\ of\ differences\ squared}{Number\ in\ sample} = \frac{31.215}{13} = 2.401$$

Standard deviation σ:

$$\sqrt{S} = \text{sigma } \sigma = \sqrt{2.401} = 1.549$$

GROUPED DATA

You are carrying out a quality check on suppliers. The table gives the size in mm of samples of 100 mm bottom bracket spindles.

Line	Range/group mm	Mid-point x	Frequency f	fx	$\bar{x}$	$(x-\bar{x})$	$f(x-\bar{x})^2$
1	89–91	90	17	1530	99.07	−9.07	1398.5
2	91–93	92	18	1656	99.07	−7.07	899.7
3	93–95	94	19	1786	99.07	−5.07	488.4
4	95–97	96	20	1920	99.07	−3.07	188.5
5	97–99	98	30	2940	99.07	−0.07	34.3
6	99–101	100	50	5000	99.07	0.93	4326
7	101–103	102	35	3570	99.07	2.93	300
8	103–105	104	22	2288	99.07	4.93	535
9	105–107	106	18	1908	99.07	6.93	864
10	107–109	108	10	1080	99.07	8.93	89
			Total: 239	Total: 23,678			Total: 9,123.4

Mean:

$$\bar{x} = \frac{\sum(fx)}{\sum f} = \frac{23678}{239} = 99.07$$

Variance:

$$S = \frac{f(x - \bar{x})2}{\sum f} = \frac{9123.4}{239} = 38.17$$

Standard deviation σ:

$$\text{Sigma } \sigma = \sqrt{S} = \sqrt{38.17} = 6.178$$

Sales analysis spreadsheet.

PROFIT AND LOSS ACCOUNT

This is essential data to see if the company is making a profit or a loss. With the growth of Community Interest Companies—CICs, generally not-for-profit organizations, these need to be crystal clear to show all incomes and outgoings. This is essential to assure the HMRC and others where the money is going.

ANY CYCLE COMPANY

Profit and Loss Account

Income £

Bicycle sales*: 25,000

Accessory sales*: 4,000

Service work: 3,000

Repair work: 10,000

*Less purchase costs

Total: 42,000

Expenses £

Rent of premises: 5,000

Business rates: 1,000

Electricity: 250

Water: 300

Gas: 300

Transport: 1,500

Technology: 250

Total: 8,600

Trading Profit: £33,400

BALANCE SHEET

The balance sheet sets out the company's position with the relationship between what it owns and what it owes at any one point in time, to keep abreast of the trading it is normal to produce a balance sheet for each quarter of the year and compare to previous similar quarters. The bicycle industry has variations with the seasons, so each quarter results may be very different, but the comparison of any quarter with those of the previous year's same quarter should give a good indication of the trading situation.

Any cycle company		
Balance sheet		
	2019	2020
	£	£
Current assets		
Cash in hand	1500	500
Cash in bank	10500	13000

(Continued)

Any cycle company		
Balance sheet		
	2019	2020
	£	£
Stock at valuation	120000	130000
30-day assets	20000	15000
Inventory assets	10000	12000
Subtotal	16200	170500
Long-term assets		
Tools and equipment	10000	7000
Goodwill	2000	5000
Premises	30000	40000
Subtotal	**42000**	**52000**
Total current assets	**204000**	**225000**
Current liabilities		
Bills payable	30000	32000
Loans payable	10000	8000
Tax payable	10000	10000
Total current liabilities	**50000**	**50000**

Acid-Test Ratio (ATR)—This is the absolute test of a company's viability. It is calculated by dividing the company's total current assets (TCA) by the total current liabilities (TCL).

$$ATR = TCA / TCL$$

For the two years shown in the balance sheet these are:

2019 ATR = 204000 / 50000 = 4.08

2020 ATR = 222500 / 50000 = 4.45

It can be seen that the ATR has improved as the company has matured. If the ATR is greater than unity (1) the company is viable; if less then unity there is likely to be a need to borrow money.

SOME EXAMPLES OF DATA YOU MIGHT COLLECT

Event timing, transponder data, personal data, heart rate, cadence, gear ratio, Vo2 max

Chapter 13

Health, safety, security and the environment

Health, Safety and the Environment are controlled by a number of Acts of Parliament and subsequent regulations and statutory instruments. These may have regional variations, or specifics relating to the bicycle and related associated industries. These topics are also impacted by other laws, for instance, those relating to the countryside. Perhaps the most important aspect of the Health and Safety laws is that they are vicarious. That is, if you are a manager, or supervisor, of a bicycle factory, you may be prosecuted for any injury or death caused by one of the technicians, or other staff actions, as well as the staff, involved being prosecuted. Breaking Health and Safety, or Environmental Laws may result in custodial sentences as well as fines and damages to the injured parties. Meeting the requirements of the Health, Safety and Environmental laws and regulations is the responsibility of everybody—ignorance of the law is not an excuse, so you need to take care.

PERSONAL HEALTH AND SAFETY PROCEDURES

Skincare (personal hygiene) systems

All employees should be aware of the importance of personal hygiene and should follow correct procedures to clean and protect their skin in order to avoid irritants causing skin infections and dermatitis. All personnel should use a suitable barrier cream before starting work and again when recommencing work after a break. There are waterless hand cleaners available which will remove heavy dirt on the skin prior to thorough washing. When the skin has been washed, after-work restorative creams will help to restore the skin's natural moisture.

Many paints, refinishing chemicals and bicycle repair shop materials will cause irritation on contact with the skin and must be removed promptly with a suitable cleansing material. Paint solvents may cause dermatitis, particularly where the skin has been in contact with peroxide hardeners or acid catalysts: these have a drying effect which removes the natural oils in the skin. There are specialist products available for the paint shop which will remove these types of materials from the skin quickly, safely and effectively.

Hand protection

Bicycle technicians are constantly handling substances which are harmful to health. The harmful effect of liquids, chemicals and materials on the hands can be prevented, in many cases, by wearing the correct type of gloves. To comply with Control of Substances Hazardous to Health (COSHH) Regulations, vinyl disposable gloves must be used when refinishing paintwork to give skin protection against toxic substances. Other specialist gloves available are rubber and polyvinyl chloride gloves for protection against solvents, oil and acids; leather gloves for hard-wearing and general repair work in the bike shop; and welding gauntlets, which are made from specially treated leather and are longer than normal gloves to give adequate protection to the welder's forearms.

Protective clothing

It is worn to protect the worker and his/her clothes from coming into contact with dirt, extremes of temperature, falling objects and chemical substances. The most common form of protective clothing is the overall a one-piece boiler suit made from good quality cotton, preferably flame-proof. Nylon and other synthetic materials tend to be highly flammable and therefore pose a hazard in the vicinity of naked flames. Worn and torn materials should be avoided as they can catch in moving machinery. Where it is necessary to protect the skin, closely fitted sleeves should be worn down to the wrist with the cuffs fastened. All overall buttons must be kept fastened, and any loose items such as ties and scarves should not be worn. The coveralls must withstand continuous exposure to a variety of chemicals. They can be of the one-piece variety or can have separate disposable hoods

Eye protection

It is required when there is a possibility of eye injury from flying particles when using a grinder disc sander, power drill or pneumatic chisel. Many employers are now requiring all employees to wear some form of safety glasses when they are in any workshop area. There is always the possibility of flying objects, dust particles, or splashing liquids entering the eyes. Not only is this painful but it can, in extreme cases, cause loss of sight. Eyes are irreplaceable: therefore, it is advisable to wear safety goggles, glasses or face shields in all working areas.

Foot protection

Safety footwear is essential in the workshop environment. Boots or shoes with steel toecaps will protect the toes from falling objects. Rubber boots will give protection from acids or in wet conditions. Never wear defective footwear as this becomes a hazard in any workshop environment.

Ear protection

The **Noise at Work Regulations 2005** defines three action levels for exposure to noise at work:

- Daily personal exposure of up to 80 dB. Where exposure exceeds this level, suitable hearing protection must be provided on request
- Daily personal exposure of up to 85 dB. Above this second level of provision, hearing protection is mandatory.
- A peak sound pressure of 87 dB.

Tech note

Noise is measured in decibels (dB)—this is measured on a scale based on logarithms. That is to say that increases do not follow the normal arithmetic scale in terms of increase in noise. An increase of 3 dB, from say 84 dB to 87 dB will give a doubling of the noise heard. So to cut the reading by 3 dB will reduce the noise heard by half

---

Fire precautions

What is fire? Fire is a chemical reaction called combustion (usually oxidation resulting in the release of heat and light). To initiate and maintain this chemical reaction, or in other words for an outbreak of fire to occur and continue, the following elements are essential:

Fuel: A combination substance, solid, liquid or gas.

Oxygen: Usually air, which contains 21% oxygen.

Heat: The attainment of a certain temperature—once a fire has started it normally maintains its heat supply.

Methods of extinction of fire: Because three ingredients are necessary for fire to occur, it follows logically that if one or more of these ingredients are removed, fire will be extinguished. Basically, three methods are employed to extinguish a fire: removal of heat (cooling); removal of fuel (starving); and removal or limitation of oxygen (blanketing or smothering).

Removal of heat: If the rate of heat generation is less than the rate of dissipation, combustion cannot continue. For example, if cooling water can absorb heat to a point where more heat is being absorbed than generated, the fire will go out.

Removal of fuel: This is not a method that can be applied to fire extinguishers. The subdividing of risks can starve a fire, prevent large losses and enable portable extinguishers to retain control; for example, part of a building may be demolished to provide a fire stop. The following advice can contribute to a company's fire protection programme:

What can cause fire in this location and how can it be prevented?

- If fire starts, regardless of cause, can it spread?
- If so, where to?
- Can anything be divided or moved to prevent such spread?

   Removal or limitation of oxygen: It is not necessary to prevent the contact of oxygen with the heated fuel to achieve extinguishment. It will be found that where most flammable liquids are concerned, reducing the oxygen in the air from 21 to 15% or less will extinguish the fire. Combustion becomes impossible even though a considerable proportion of oxygen remains in the atmosphere. This rule applies to most solid fuels although the degree to which oxygen content must be reduced may vary. Where solid materials are involved they may continue to burn or smoulder until the oxygen in the air is reduced to 6%. There are also substances which carry within their structures sufficient oxygen to sustain combustion.

   Fire risks in the workshop: Fire risks in the repair shop cover all classes of fire: class A is paper, wood and cloth; class B is flammable liquids such as oils, spirits, alcohols, solvents and grease; class C is flammable gases such as acetylene, propane, butane; and also, electrical risks. It is essential that fire is detected and extinguished in the early stages. Workshop staff must know the risks involved and should be aware of the procedures necessary to combat a fire. Workshop personnel should be aware of the various classes of fire and how they relate to common workshop practice.

   Class A fires; wood, paper and cloth: These fires often start by carelessness, throwing hot or burning objects in the waste bin—always use waste bins with lids which will prevent the spread of fires.

   Class B fires; flammable liquids: Flammable liquids are the stock materials used in the trade for all bicycle refinishing processes: a gun cleaner to clear finish coats, cellulose to the more modern finishes, can all burn and produce acrid smoke.

   Class C fires; gases: Lots of aerosol products contain flammable gasses, these can be ignited by sunlight on hot days.

   Electrical hazards: Electricity is not of itself a class of fire. It is, however, a potential source of ignition for all of the fire classes mentioned above. The Electricity at Work Regulations covers the care of cables, plugs and wiring. In addition, in the workshop, the use of welding and cutting equipment produces sparks which can, in the absence of good housekeeping, start a big fire. Training in how to use firefighting equipment can stop a fire in its early stages. Another hazard is the electrical energy present in all large batteries. A short-circuit across the terminals of a battery can produce sufficient energy to form a weld and in turn heating, a prime source of ignition.

General precautions to reduce fire risk

- Good housekeeping means putting rubbish away rather than letting it accumulate.
- Read the manufacturer's material safety data sheets so that the dangers of flammable liquids are known.
- Only take from the stores sufficient flammable material for the job in hand.
- Materials left over from a specific job should be put back into a labelled container so that not only you but anyone (and this may be fire personnel) can tell what the potential risk may be.
- Be extremely careful when working close to plastic components.
- Petrol tanks are a potential hazard: supposedly empty tanks may be full of vapour. To give some idea of the potential problem, consider 5 litres (1 gallon) of petrol: it will evaporate into 1 m^3 (35 feet^3) of neat vapour, which will mix with air to form 14 m^3 (500 feet^3) of flammable vapour. Thus the average petrol tank needs only a small amount of petrol to give a tank full of vapour.

The keys to fire safety are:

- Take care.
- Think.
- Train staff in the correct procedures before things go wrong.
- Ensure that these procedures are written down, understood and followed by all personnel within the workshop.

Tech note

Any carbon-based material will burn in air if at the temperature needed for combustion—be aware of this with dust in the factory or workshop.

Types of portable fire extinguishers

Water is the most widely used extinguisher agent. With portable extinguishers, a limited quantity of water can be expelled under pressure and its direction controlled by a nozzle.

There are basically two types of water extinguishers. The gas (CO_2) cartridge operated extinguisher, when pierced by a plunger, pressurizes the body of the extinguisher, thus expelling the water and producing a powerful jet capable of rapidly extinguishing class A fires. In stored pressure extinguishers, the main body is constantly under pressure from dry air or nitrogen, and the extinguisher is operated by opening the squeeze grip discharge valve. These extinguishers are available with 6 or 9 litres capacity bodies and thus provide alternatives to weight and accessibility.

Foam is an agent most suitable for dealing with flammable liquid fires. Foam is produced when a solution of foam liquid and water is expelled under pressure through a foam-making branch pipe at which point air is entrained, converting the solution into the foam. Foam extinguishers can be pressurized either by a CO_2 gas cartridge or by stored pressure. The standard capacities are 6 and 9 litres.

Spray foam. Unlike conventional foams, aqueous film-forming foam (AFFF) does not require to be fully aspirated in order to extinguish fires. Spray foam extinguishers expel an AFFF solution in an atomized form, which is suitable for use on class A and class B fires. AFFF is a fast and powerful means of tackling fire and seals the surfaces of the material, preventing re-ignition. The capacity can be 6 or 9 litres, and operation can be by CO_2 cartridge or stored pressure.

Carbon dioxide

Designed specifically to deal with class B, class C and electrical fire risks, these extinguishers deliver a powerful concentration of carbon dioxide gas under great pressure. This not only smothers the fire very rapidly but is also non-toxic and is harmless to most delicate mechanisms and materials.

Dry powder

This type of extinguisher is highly effective against flammable gases, open or running fires involving flammable liquids such as oils, spirits, alcohols, solvents and waxes and electrical risks. The powder is contained in the metal body of the extinguisher from which it is supplied either by a sealed gas cartridge or by dry air or nitrogen stored under pressure in the body of the extinguisher in contact with the powder.

Dry powder extinguishers are usually made in sizes containing 1 to 9 kg of either standard powder or (preferably and more generally) all-purpose powder, which is suitable for mixed risk areas.

Safety signs in the workshop—it is a legal requirement that all safety signs comply with BS EN ISO 7010:2012+A7:2017.

Prohibition signs have a red circular outline and crossbar running from top left to bottom right on a white background. The symbol displayed on the sign must be black and placed centrally on the background, without obliterating the crossbar. The colour red is associated with 'stop' or 'do not'.

Warning signs have a yellow triangle with a black outline. The symbol or text used on the sign must be black and placed centrally on the background. This combination of black and yellow identifies caution.

Mandatory signs have a blue circular background. The symbol or text used must be white and placed centrally on the background. Mandatory signs indicate that a specific course of action is to be taken.

Safe condition signs provide information for a particular facility and have a green square or rectangular background to accommodate the symbol or text, which must be in white. The safety colour green indicates 'access' or 'permission'.

General safety precautions in the workshop

The Health and Safety Act is designed to ensure that:

- Employers provide a safe working environment with safety equipment and appropriate training.
- Employees work in a safe manner using the equipment provided and follow the guidance and training which is provided.
- Customers and other entering any business premises are safe and protected.

REMEMBER

It is all about keeping yourself, your colleagues and your customers safe, as you would want them to keep you safe too.

Particular hazards may be encountered in the manufacturing or repair of bicycles, and safety precautions associated with them are as follows:

1. Do wash before eating, drinking or using toilet facilities to avoid transferring the residues of sealers, pigments, solvents, filing of steel, lead and other metals from the hands to the inner parts and other sensitive areas of the body.
2. Do not use kerosene, thinners or solvents to wash the skin. They remove the skin's natural protective oils and can cause dryness and irritation or have serious toxic effects.
3. Do not overuse waterless hand cleaners, soaps or detergents, as they can remove the skin's protective barrier oils.
4. Always use a barrier cream to protect the hands, especially against oils and greases.
5. Do follow work practices that minimize the contact of exposed skin and the length of time liquids or substances stay on the skin.
6. Do thoroughly wash contaminants such as dirty oil from the skin as soon as possible with soap and water. A waterless hand cleaner can be used when soap and water are not available. Always apply skin cream after using waterless hand cleaner.
7. Do not put contaminated or oily rags in pockets or tuck them under a belt, as this can cause continuous skin contact.

8. Do not dispose of dangerous fluids by pouring them on the ground, or down drains or sewers.

9. Do not continue to wear overalls which have become badly soiled or which have acid, oil, grease, fuel or toxic solvents spilt over them. The effect of prolonged contact from heavily soiled overalls with the skin can be cumulative and life-threatening. If the soilants are or become flammable from the effect of body temperature, a spark from welding or grinding could envelop the wearer in flames with disastrous consequences.

10. Do not clean dusty overalls with an airline: it is more likely to blow the dust into the skin, with possible serious or even fatal results.

11. Do wash contaminated or oily clothing before wearing it again.

12. Do discard contaminated shoes.

13. Wear only shoes which afford adequate protection to the feet from the effect of dropping tools and sharp and/or heavy objects on them, and also from red hot and burning materials. Sharp or hot objects could easily penetrate unsuitable footwear such as canvas plimsolls or trainers. The soles of the shoes should also be maintained in good condition to guard against upward penetration by sharp or hot pieces of metal.

14. Ensure gloves are free from holes and are clean on the inside. Always wear them when handling materials of a hazardous or toxic nature.

15. Keep goggles clean and in good condition. The front of the glasses or eyepieces can become obscured by welding spatter adhering to them. Renew the glass or goggles as necessary. Never use goggles with cracked glasses.

16. Always wear goggles when using a bench grindstone or portable grinders, disc sanders, power saws and chisels.

17. When welding, always wear adequate eye protection for the process being used. MIG/MAG welding is particularly high in ultraviolet radiation which can seriously affect the eyes.

18. Glasses, when worn, should have 'safety' or 'splinter-proof' glass or plastic lenses.

19. Always keep a suitable mask for use when dry flatting or working in dusty environments and when spraying adhesive, sealers, solvent carried waxes and paints.

20. In particularly hostile environments such as when using volatile solvents or isocyanate materials, respirators or fresh air fed masks must be worn.

21. Electric shock can result from the use of faulty and poorly maintained electrical equipment or misuse of equipment. All electrical equipment must be frequently checked and maintained in good condition. Flexes, cables and plugs must not be frayed, cracked, cut or damaged in any way. Equipment must be protected by the correctly rated fuse.

22. Use low-voltage equipment wherever possible (110 volts).
23. In case of electric shock:
    a. Avoid physical contact with the victim.
    b. Switch off the electricity.
    c. If this is not possible, drag or push the victim away from the source of the electricity using a non-conductive material.
    d. Commence resuscitation if trained to do so.
    e. Summon medical assistance as soon as possible.

Electrical hazards

The Electricity at Work Act 1989 fully covers the responsibilities of both the employee and the employer. You are obliged to follow these regulations for the protection of yourself and your colleagues. Some of the important points to be aware of are given below.

Voltages—the normal mains electricity voltage via a three-pin socket-outlet is 240 volts; heavy-duty equipment such as machine tools uses 415 volts in the form of a three-phase supply. Both 240 volt and 415 volt supplies are likely to kill anybody who touches them. Supplies of all voltages must be used through a professionally installed system and be tested regularly. If 240 volts is used for power tools, then a safety circuit breaker should be used. A safer supply for power tools is 110 volts; this may be wired into the workshop as a separate circuit or provided through a safety transformer. Inspection hand-lamps are safest with a 12-volt supply.

Checklist

Before using electrical equipment you are advised to check the following:

1. **Cable condition:** Check for fraying, cuts or bare wires.
2. **Fuse rating:** The fuse rating should be correct for the purpose as recommended by the equipment manufacturer.
3. **Earth connection:** All power tools must have sound earth connections.
4. Plugs and sockets—do not overload plugs and sockets; ensure that only one plug is used in one socket.
5. **Water:** Do not use any electrical equipment in any wet conditions.
6. **PAT testing:** It is a requirement of the Electricity at Work Regulations that all portable electrical appliances are tested regularly, they should be marked with approved stickers and the inspection recorded in a log.

Control of Substances Hazardous to Health (COSHH)

COSHH regulations require that assessments are made of all substances used in the workshop. This assessment must state the hazards of using the materials and how to deal with accidents arising from misuse. Your wholesale supplier will provide you with this information as set out by the manufacturer in the form of either single sheets on individual substances, or a small booklet covering all the products in a range. Datasheets are available for many common products on the various manufactures and suppliers websites. An example can be found in the Appendix.

Reporting of Injuries, Diseases and Dangerous Occurrences Regulations (RIDDOR)

RIDDOR 2013 requires that certain information is reported to the Health and Safety Executive (HSE). This includes the following:

- Work-related accidents which cause deaths.
- Work-related accidents which cause certain serious injuries (reportable injuries).
- Diagnosed cases of certain industrial diseases.
- Certain 'dangerous occurrences' (incidents with the potential to cause harm).

Maintain the health, safety and security of the work environment

It is the duty of every employee and employer to comply with the statutory regulations relating to health and safety and the associated guidelines, which are issued by the various government offices. That means you must work in a safe and sensible manner. You are expected to follow the health and safety recommendations of your employer; employers are expected to provide a safe working environment and advise on suitable safe working methods.

Health and safety law states that organizations must:

- Provide a written health and safety policy (if they employ five or more people).
- Assess risks to employees, customers, partners and any other people who could be affected by their activities.
- Arrange for the effective planning, organization, control, monitoring and review of preventive and protective measures.
- Ensure they have access to competent health and safety advice.
- Consult employees about their risks at work and current preventive and protective measures.

Tech note

Everybody in an organization has a duty of care related to health and safety. The HSE may bring about prosecutions, or lesser prohibitions subject to timed actions—for instance, being given a short period of time to rectify a machine fault. However, the final consequences can be devastating for a firm and its employees, possible outcomes are:

- Unlimited fine
- Imprisonment
- Closing down of the business
- Disqualification from working in that job or type of business

---

You must work in a safe manner, or you are breaking the Health and Safety at Work Act and are liable to a fine, and possible imprisonment and maybe disqualification from working in that job. This means following the safe working practices which are normally used within the industry. The guidelines published by the HSE and textbooks usually identify industry-accepted safe working practices. Examples of important procedures are:

- Always the correct support for heavy objects.
- Always wear overalls, safety boots and any other personal protective equipment (PPE) when it is needed, for example, safety goggles when grinding or drilling and a breathing mask when working in dusty conditions.
- Always use the correct tools for the job.

Where identified hazards cannot be removed, appropriate action is taken immediately to minimize risk to own and others' health and safety

This section is about those situations where the hazards cannot be readily removed, that is, how do you behave in accident situations, or when equipment malfunctions and you can see an accident about to happen?

Dangerous situations are reported immediately and accurately to authorized persons

As a trainee in the bicycle industry, your company will require you to report any dangerous situations to your supervisor; this will be a person that you know as the charge-hand, foreperson or service manager. Any internal matter should in the first instance be reported to one of these people—you will know who this is from your induction training. However, if you are working alone or the matter is not a company one, then you must inform the relevant authority. The four emergency services in the UK are Police, Fire, Ambulance and Coast Guard. To call them, use any telephone and dial 999.

Suppliers' and manufacturers' instructions relating to safety and safe use of all equipment are followed

Many pieces of equipment are marked 'only to be used by authorized personnel'. This is mainly because incorrect use can cause damage to the equipment, the work-piece or the operator. Do not operate equipment which you have not been properly trained to use and have not been given specific permission to use.

The suppliers of garage equipment issue operating instructions and as part of your training you must read these instruction booklets so that you will understand the job better. You will also find that certain safety instructions are marked on the equipment. Safe Working Load (SWL) in either tonnes or kilograms is marked on lifting equipment. You must ensure that you do not exceed these maximum load figures. Some items of equipment have two-handed controls or dead-man grips—do not attempt to operate these items incorrectly.

Approved/safe methods and techniques are used when lifting and handling

Do not attempt to manually carry a load which you cannot easily lift and which you cannot see above and around. The advised maximum weight of the load that you should lift is 20 kilograms, but as a trainee, this may still be too heavy for you. Do not lift weights that you are not comfortable with.

When you are lifting items from the floor always keep your back straight and bend your knees. Bending your back whilst lifting can cause back injury. If you keep your feet slightly apart, this will improve your balance. It is always a good idea to wear safety gloves when manually lifting.

Required personal protective clothing and equipment are worn for designated activities and in designated areas

The following table lists typical items of PPE and states when they must be worn.

You will often see safety notices requiring you to wear certain PPE in some areas at all times, this is because other people are working in the area and you may be at risk.

Injuries involving individuals are reported immediately to competent first aiders and/or appropriate authorized persons and appropriate interim support is organized to minimize further injury

Should there be an accident the first thing to do is call for help. Either contact your supervisor or a known first aid person. Should any of these not be available, and it is felt appropriate, call for your local doctor or an ambulance.

You are not expected to be a first aid expert, nor are you advised to attempt to give first aid unless you are properly qualified. However, as a professional in the bicycle industry you should be able to preserve the scene, that is, prevent further injury and make the injured person comfortable. The following points are suggested as ones worth remembering:

1. Switch off any power source.
2. Do not move the person if an injury to the back or neck is suspected.
3. In the case of electric shock turn off the electricity supply.
4. In the case of a gas leak, turn off the gas supply.
5. Do not give the person any drink or food, especially alcohol, in case surgery is needed.
6. Keep the person warm with a blanket or coat.
7. If a wound is bleeding heavily, apply pressure to the wound with a clean bandage to reduce the loss of blood.
8. If a limb has been trapped, use a safe jack to free the limb.

Visitors are alerted to potential hazards

The best policy is not to let anyone other than staff into the workshop; if entry is necessary by others then you must highlight any hazards and ensure that the company policy is complied with—usually, this is done by requiring a sign-in and issuing a visitor lanyard with safety instructions.

It is always a good idea to accompany customers when they are in the workshop, this way you can advise them in the event that they may do something potentially dangerous or if there is a hazard of which they may not be readily aware.

Injuries resulting from accidents or emergencies are reported immediately to a competent first aider or appropriate authority

If a person is injured the first action must be to ensure that first aid is given by a competent first aider or other suitable person. Most companies have a designated first aider who is trained to deal with accidents and emergencies. If your company has no such a person on the staff then you will have a designated person whom you must contact in the event of a colleague being injured. That person may be your supervisor or another senior member of the staff. If no manager or other senior person is available, you should either dial 999 for an ambulance or telephone your company doctor, then inform the garage manager.

Incidents and accidents are reported in an accident book

By law, all companies are required to maintain records of accidents which take place at work. These records are usually kept in an accident book. Accident books may be inspected by HSE inspectors they must be kept for a period of at least three years from the date of the last entry.

The information which is required to be recorded in the accident book is:

- Name and address of the injured person.
- Date, time and place of accident/dangerous occurrence.
- Name of person making the report and date of entry.
- A brief account of the accident and details of any equipment/substances which were involved.

It is always a good idea to keep a notepad to help remind you which way round things go when working on unfamiliar bicycles, or with new machinery, this would also be useful for making any other notes, such as those about an accident.

Where there is a conflict over limitation of damage, priority is always given to the person's safety

You can always buy a new bicycle, but you cannot buy a new arm for a mechanic. In the event of an accident, people come first. For instance, if a building is on fire, do not re-enter to retrieve your belongings, wait until the fire is out and there is no risk before going back into the building.

NOTES:

- Health and safety issues are further discussed in the appropriate chapters.
- Examples of a risk Assessment and a COSHH Data Sheet can be found in Appendix I.

MISCELLANEOUS TOPICS

Lone working: Lone working is quite common in the retail and repair side of the bicycle industry. There should be a statement in the company's health and safety policy to address this topic. A very small accident could lead to death or serious injury if the lone worker is unable to quickly call help. Keeping a mobile phone readily accessible is good practice with an emergency contact number on speed-dial.

First aid: The HSE recommends that in a company with 5–50 workers there should be at least one trained first-aider, and there should be another for every 50 after that. There are different levels of first-aid training. St John Ambulance is a charity that offers first-aid training and other training and support including online advice. They train more than 4,00,000 people every year.

Security: Security is a major issue with the high value and the easy disposability of bicycles. Over 20,000 bicycles are stolen in London every year. The Metropolitan Police and many other Police forces have staff just dealing with this issue. There are also issues relating to personal security and the security of premises with the ever-growing problems related to drug-culture. Most Police forces offer online advice and guidance on security and crime prevention.

Car bicycle racks: With the growth of the popularity of cycling, and the tendency to take bicycles on holiday, there is a growing use of bicycle racks. These may be on the car roof or attached to the rear of the vehicle. Experience shows that both bicycles and racks can easily become detached from the vehicle. The author is informed by traffic officers that on many stretches of UK motorway it is normal to have one or two bicycles become detached every week. These, of course, may cause serious accidents involving other vehicles. These can lead to complicated and expensive insurance claims. The best place for a bicycle is inside the vehicle. With the wheels, removed bicycles can usually be fitted inside even very small cars.

Anti-theft security: There are many anti-theft security systems. There is no absolute way of preventing theft, but security marking and fitting security chips may help to get a stolen bicycle returned if found by the Police. During raids on criminal suspects, many high-value objects are found each day. If yours is marked it will lead to its return, this may also help the conviction of the criminal by providing additional evidence.

Emergency procedures: All businesses should have emergency procedures written into their policy documents.

Emergency contacts: All businesses are advised to maintain lists of emergency contacts, a list situated near the telephone and a number next to the intruder alarm. Cycling clubs and event organisers are required by the governing bodies to keep emergency contact details for all participants.

Club riding: Club riding requires a disciplined approach to riding; it only comes with practice. Most cycling clubs have different levels of groups according to speed and distance. They will usually advise you to start in the slowest and shortest distance group; in this group, the leaders will instruct you in save riding procedures. As you learn the technique of group riding your speed and endurance will increase enabling you, if you wish, to move to another group.

Lone riding: Audax and other similar events often lead to lone riding. If it is an organised event then the organisers will keep rider details and emergency contact details. If you are touring alone you are advised to tell someone the area that you are visiting and have some form of a daily check-in procedure. Especially with the increasing number of cyclists visiting far-flung places and the increase in random war zones—please keep in touch with someone. Please remember, many cultures do not understand why anybody would want to cycle anywhere, so they may see cyclists as enemies.

Insurance: Most cycling clubs offer insurance for cyclists. Basic club membership often covers third-party insurance and legal cover. This is a basic minimum. So that you are covered for damage caused to others and have a solicitor to fight your case should you be involved in an accident. The three national cycling clubs in the UK that offer this as part of the membership package are:

- National Clarion Cycling Club—which has sections across the UK and in Italy.
- Cycling UK—formerly called the CTC—which has sections across the UK, often called CTC followed by the local town name.
- British Cycling—formerly called the BCF. Most cycling clubs are affiliated to British Cycling and offer British Cycling and insurance as an add-on.

Most other countries have a similar structure of clubs and groups, although they are often more related to racing and other similar activities. In the USA, it is almost 100% of cyclists that have insurance cover, either through a club or personally.

Chapter 14

The bicycle industry

The bicycle industry, like many other industries, is going through a large-scale change; but this is not new, the bicycle is constantly being developed and hence the ways of making them and selling them is constantly changing too. There is a saying that nothing is constant except change—there are likely to be disruptions to the bicycle industry for as long as people ride bicycles and others want to make a living from them. This chapter looks at the industry past, present and the possible future.

I hope that you may read this chapter and gain an insight which you will use to start a new business, or to improve a current one, so as to keep the bicycle industry alive for future generations to enjoy.

TRADITIONAL BICYCLE SHOP

Bicycle retailing as traditionally been through the bicycle shop serving a local area. The local bicycle shop would supply a range from children's machines to those of serious club riders. It was normal for customers to go to the same shop for each stage of their cycling life, that is, from starting with a children's kiddy bike, going through the intermediate stages to a serious adult cycle. In the meantime, the cycle shop provided maintenance and accessories. This was the same business model as that of the local garage and gas (petrol) station, selling and repairing cars alongside petrol and other consumables.

Bicycle shops using that model have been in existence since the time that bicycling became popular—about 1890. Some current businesses have existed since then going through different phases of retailing, small scale manufacturing and selling other items such as mechanical toys, small motorcycles and even small boats. The traditional bicycle industry in the UK was decimated in the 1960s when cars became more affordable, it flourished again when the Olympics of 2012 saw great success to the GB Cycling team. In other countries, the bicycle business model has been, indeed currently is, very different.

Perhaps the reason for the different business models is the actual position of the bicycle in society and its usefulness for making money. In the UK, the bicycle as always had a hobby or enthusiast status. Nobody has ever really needed a bicycle to do their job in the UK. More people than ever use them for commuting to work in the UK cities than ever before, but nobody actually needs one to commute to work as all cities have adequate public transport in the form of buses and railways and in some cases underground systems. In other countries, a bicycle is an essential form of transport, an all-embracing workhorse without which people would not be able to earn a living. In the Northern Hemisphere, most countries have a state-provided support system for people—unemployment benefit. Such systems do not exist in many other countries, particularly in the Southern Hemisphere, or the value of them is not sufficient to support a family in any decent manner. So, the bicycle is more of a tool to do a job in many countries. Hence the bicycle business model is different. Bicycles are sold and repaired as economically as possible, and in a way to meet the needs of the people.

This means that in areas where people live and work there will be a bicycle business to meet specific needs.

There is also a change in the way in which goods and services are provided; this appears to be related to a number of different issues—this seems to apply across other areas than just bicycles and across the world too.

MANUFACTURING

Bicycle manufacturing, like any other manufacturing, is about return on capital investment per unit made. The more units that can be made for the given capital investment will spread the cost further allowing lower prices and greater profits. Let's briefly look at this.

If you are setting up manufacturing bicycles you have three sets of costs to consider, these are:

1. Initial capital costs
2. Fixed running costs
3. Variable manufacturing costs

This model also applies to retail organizations as well, in fact almost any business.

Initial capital costs

The amount of funding needed to set up the business, this includes:

- **The building:** For manufacturing, this usually entails some form of specialist construction for delivery and loading areas, ventilation for the specialist processes such as welding and paint spraying, high level of electrical and other power supplies.

- **The tools and machinery:** Jigs, welding equipment, bending and forming tools, paint/finishing equipment, drilling and other machining equipment.
- **Operating funds:** The money needed to pay for initial materials, labour costs and up-front charges such as rent, business rates, energy supplies and advertising/marketing.
- **Intellectual property:** This can be the most valuable item, the designs and ideas for the bicycle. For the guys at Marin County who came up with the design of the initial mountain bike this was their winning investment—it game-changed the bicycle industry.

One of the problems with the cycle industry is the lack of investment in manufacturing. In fact, this applies across all manufacturing. Manufacturing needs large investments which do not see immediate returns—they have to be made for the long-game. Also, the costs of borrowing money need to be taken into account—borrowing is both expensive and can be risky if repayment dates are not met. Our industry needs bicycling enthusiasts who can see the long-term benefits of manufacturing bicycles.

Fixed costs

These are the costs which need to be covered each month, irrespective of the number of bicycles that are made. These will include building costs—rent, business rates, fixed energy costs, maintenance and fixed staff costs.

Variable costs

These are the costs to make each bicycle, mainly materials and direct labour. As the output of bicycles increases the variable costs will increase too. Of course, if increased production is because of increased sales the revenue will increase too if the pricing is constant. The management challenge in running any manufacturing business is the balancing of costs and sales to ensure passing the **break-even point**, that is, where the revenue exceeds both the fixed costs and the variable costs and the business is making a surplus, in other words, a profit.

Sometimes companies have rapid growth, expand and so increase costs, then a slump in sales and the company is out of business because it cannot meet its costs. This happens in the cycle industry sometimes, but many old companies have kept steady by controlling growth and maintaining steady production. A well-documented example from the automotive industry is the Morgan Motor Company; they manufacture a range of sports cars. Morgan has been making cars for over 110 years, their production rate remains fairly constant, but their waiting list time changes. As I write this the Morgan waiting list is six months, it has been as long as ten years. It shows that if you make a good product, customers will wait for delivery.

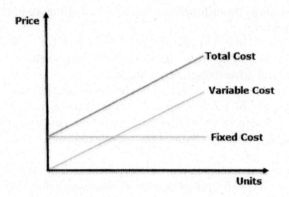

Figure 14.1 Break-even graph showing where production moves into profit.

PRODUCTION PLANNING

Aircraft producer Boeing has a rolling 3-year production schedule for their works in Seattle. Customers are advised of this when placing orders, but some changes to specification can be made before production starts. Some bicycle manufactures also have similar production schedules.

Mass production

It is about producing large numbers of identical components. Mass production originally started in about 1861 at the start of the American Civil War. The war saw a large number of guns being made to satisfy both Federate and Confederate armies, the gun makers developed mass production method, these methods were then used in the production of other items, which of course included bicycles.

The key points of mass production are:

- A large quantity of identical parts is used.
- Each person specializes in doing only one part of the manufacturing process.
- Components are designed for ease of assembly.
- A flow line, or an assembly line, is used to move the product between the different stages of manufacture.

This tends to lead to lots of identical bicycles being made and a limited variation in frame sizes. Bicycles produced in this way tend only to be available in sizes of small, medium and large. They also usually have a very limited paint colour range too. When you consider the numbers of bicycles being made, it is easy to see why the ranges are limited and the same components are used throughout the range. Bicycle manufacturers using mass production methods will often produce up to 10,000 bicycles per week for worldwide distribution. One large UK retail bicycle outlet which has branches in most major towns sells over 20,000 bicycles each week, drawing its stock from a variety of sources.

Hand-made craft production

In contrast to mass production, there is the making of individual bicycles for individual customers. Every single detail can be tailored for the cyclist who is buying the bicycle. The price of these hand-made machines is about ten times that of the mass-produced ones.

There are several UK based bicycle makers that have been in business for over 100 years. The businesses produce hand-made frames and fit them with components of the buyer's choice. Some makers produce the frames on-site; others use specialist frame makers and paint sprayers. In the UK this is known as the **lightweight trade**. Frame builders usually spend their time and skills actually cutting and brazing the frames, a very specialized skill. Some frame builders make frames for several different bicycle companies. The frames are made to the specification of the company, using their choice of tubing and lugs, so they do not look similar at all.

Handmade frames are usually made from either steel or titanium, occasionally aluminium. That is materials which can be joined with the minimum of equipment. The frame is set-up in a jig and hand brazed or welded one joint at a time. Cutting the tubes to size and brazing the joints can take almost a week for a frame. TIG welding aluminium tubes is much faster. Tubes are bought as framesets of a generic size. The rider is measured for fit, like being measured for a tailor-made suit, and the tubes cut accordingly. A frame jig is used to ensure that the relative positions of the tubes, the angles, are held in place whilst the joints are made. Joints may use lugs, or maybe lugless, also called fillet brazed.

The production of handmade frame is slow and expensive, and usually takes place behind closed doors. There are a number of shows where frame builders show off their skills, in the UK the major specialist is Bespoked UK Handmade Bicycle Show, held in spring in Bristol in the West of England. In the USA the major show is the North America Hand Made Bicycle Show based in Sacramento, California.

Off-shore production

To cut costs and therefore increase profits, many cycle manufactures have moved to have their frames and components manufactured outside their home country. Production tends to be centred in the far east, that is, China and Taiwan. It is worth noting that China and Taiwan have close connections politically and economically, although there are many differences culturally.

Far Eastern manufacturers tend to offer three options:

- Cycle frames of a generic design
- The facility to customize the basic design—with your name, logo of special features
- Building to your specification

The generic designs are those that you will find of the major UK/North American brands. These may be steel—not a named steel, aluminium or carbon fibre. The customization is usually about finish and logos, these will attract an additional charge. Building to your specification is, of course, much more expensive. All orders are subject to minimum order quantities, shipping costs and taxes.

If you are looking for smaller quantities and higher quality, then several European manufactures can offer great bespoke services—especially those in Italy and Spain, countries with have great cycling heritage which is not always recognized.

Assembly methods

Bicycle assembly is a skilled job; it takes time and a high degree of skill. Accurate assembly is essential for both safety and the aesthetics of the finished bicycle. Manufacturers are constantly looking for ways to both reduce the time taken and the skill required. Time of skilled labour is very expensive.

One of the major cost items, other than the frame, are the wheels. At one time all wheels where hand-assembled, now only specialist ones are hand-assembled, wheel building machines are faster and more accurate—certainly in terms of consistency.

There is another factor to consider too, the delivery costs and the associated warehousing of bicycles is in effect part of the total price. It is common for the delivery of an item to cost a third of the retail price. To make the packaging of bicycles as small as possible manufactures usually pack them in a semi-knocked down form.

Knocked down form depends on whether the bicycle is being supplied to a retailer or mail order to a customer. The difficult and potentially safety issues are related to the following items:

- Brake cables and brake alignment against the wheels. Incorrectly aligned, or not correctly tightened brake components will certainly cause braking problems and probably a serious accident. With hydraulic disc brakes, the fault may not be immediately obvious until several applications if the hydraulic fluid leaks or is lost.

- **Wheel fitting:** Wheel fitting of course impacts on braking and gear changing as well as the risk of the wheel jumping out.
- **Gear cable adjustment:** Whether mechanical or electronic gear adjustment is a precision skill and usually needs to be re-set after a couple of weeks usage.
- Chain fitting. If using a split-link this needs to be firmly set in place. If using a chain tool, this requires practised skill.
- **Handle bars and saddle:** Two major setting and adjustments that have one hundred per cent impact on the rideability of the bicycle—most non-enthusiast cyclists do not know how to get comfortable—so they stop riding after a short time.
- Other items and accessories—some of these are complex to fit, or at least to fit in such a way that they are fully useable.

Of course, this list is not comprehensive, however, it highlights some of the problems related to assembly following delivery to a customer or retailer. Also, bicycle assembly requires a range of specialist tools equipment.

Model range and sizing

Hand-made bicycles can be made to a fraction of a millimetre in sizing for the customer; but mass-produced bicycles are usually only made in three sizes—small, medium and large. The sizing is generally related to the length of the seat tube and the top tube is proportional to this. On bicycles with a sloping top tube, the seat tube will be shorter. So, the sizing is based on an equivalent length; where it would be if the top tube didn't slope. Variation can be made by adjusting the saddle height with the seat pin and changing the handlebar stem length.

For mid-price range bicycles, it is normal to offer a mid-quality frame with a choice of components—usually in terms of group-set and wheel-set. Group-sets and wheel-sets are frequently more expensive than frames, so this is used to enlarge a manufacturer's range for different market segments. It also offers upgrade opportunities which customers enjoy.

- **Frame-set:** Main frame and forks giving a choice of size, colour and possible variation in geometry. In detail, there may be options in the fitting of gear hangers, bottle cage bosses, mud-guard eyes and luggage bosses.
- **Wheel-set:** Hubs, rims, tyres and spoking options. The hubs may form part of the group-set option.
- **Group-set:** Chain-set, front and rear gear mechanisms, brakes, gear block, hubs.
- **Finishing-set:** Handle bars, stem, saddle and seat-post, head-set and bottom bracket.

MODERN RETAILING

Retailing of bicycles used to be about the bicycle shop owner knowing his/ her customers, stocking items known to sell and ordering from catalogues non-stocked items. Your local shop would offer great service based on specialist knowledge. The in some areas, current retailing is more like 'stack them high and sell them cheap' as depicted in the novel *The Long Firm* by Jake Arnott. The concentration is on: sales volume, profit margin and delivery times. There are many forms of retailing, including:

- High street retail bicycle shops
- Outdoor equipment shops catering for bike packers
- Café style shops
- eBay/Gumtree, or similar shops
- Department stores
- Retail/business park based shops
- Mobile/market-based shops
- Shops at events and shows
- On-line and magazine advertised mail order

All these businesses operate on what is called the commercial, or commerce, model. It is about buying as cheap as possible and selling at a higher price as possible. In commerce, both buying a selling can be problematic. Generally, in the cycle industry, the problem is sourcing the right product at the right price to be able to sell at a profit after-sales costs. Cost of sales includes:

- Advertising
- Transport
- Post and packaging
- Stocking/warehousing
- Depreciation
- Losses and deterioration
- Cost of premises, rent and rates
- Staffing costs

Fashion

Like any other consumer product, the bicycle is subject to changes fashion. Up until the end of the 1960s bicycling as an activity was much a utility activity coupled with club-based leisure cycling. The 1970s saw a decline in the utility market, but great growth in the specialty market of the chopper. Raleigh sold about a million choppers during the 1970s. The BMX craze followed through the 1980s, with minority specialist sales continuing now. Then in the 1990s came the MTB which continues to sell. Following the 2012 Olympics in London the sports, or racing bicycle, has come back in fashion. Currently, electric bicycles are the raising sales star; but remember we had a fashion of fitting engines to bicycles over 100 years ago.

Figure 14.2 Bicycle and fashion clothing.

The way we live

Is going to influence future bicycle sales, with smaller houses and flats, and people with less time and practical skills, there is likely to be more emphasis on retailers to provide more services alongside their sales.

THE ELEPHANT BIKE

The Elephant Bike is a great example of a business model that promotes sustainable transport and socially conscious. It provides double good works and suggests possible business models for others in the bicycle industry. The Elephant Bike has in effect two parts, the UK arm which sources bicycles, provides specialist training, promotes cycling in the UK and provides funding to send bicycles to Malawi. In Malawi, it creates employment, promotes sustainable transport and provides funding for other charity work. The added benefit, as if this was not enough in itself, is that moving money and providing incomes adds to the income of others, especially those less well off.

ECONOMY NOTE

The purchasing of materials for the bicycle repairs and renovations, the paint, the transport and the living expenses of the people employed will all add to the incomes of those services providers. Typically, one pound (dollar) spent in this way will help generate five or six pounds (dollars) in the economy. Often local authorities, or other funding bodies, will provide start-up funding for this sort of business where it might meet a local need.

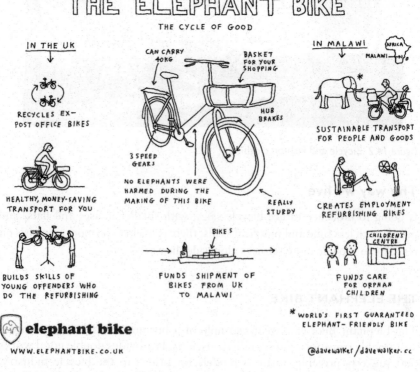

Figure 14.3 Elephant bike business concept.

Figure 14.4 Promoting the elephant bike.

Cycle-Re-Cycle is another concept for charity cycle businesses. Such companies usually trade as a CIC—Community Interest Company. These have tax advantages, but cannot register as a charity. They usually operate as not for profit companies; this allows the taking of an income and expenses. The usual basis is to source unwanted bicycles, use the process to help retrain unemployed people back into work and sell the bicycles for a small profit. All good outcomes for the community at large.

OTHER BICYCLE INDUSTRY SERVICES

Classic restorations

The growing interest in classic bicycles stems from the Eroica events—the first one being held in Italy as now spread worldwide. The same applies to the Tweed Run events which started in London and can now be sampled across the world.

The Eroica events are organised by Ciclo Club Eroica. Each year there are about 12 events across Europe, America and South Africa. As well as the actual riding of the event course there is a classification for the actual bicycle. There are set rules for the frame and components, and the rider is expected to be suitably attired too. So, around this, there is an emerging industry of suitable classic bicycles, components and clothing.

The Tweed Run events are more about having a jolly time riding a bicycle wearing Tweed clothing. Your feeding bottle rack will hold something with bubbles—Champagne for when you have finished riding may be. The events usually take place in the centre of a city, riding about 10 miles (16 km) in about 8 hours. Stopping for traditional tea and cakes in the city square, and sandwiches in the park. Rather like a day at a horse racing event such as Royal Ascot.

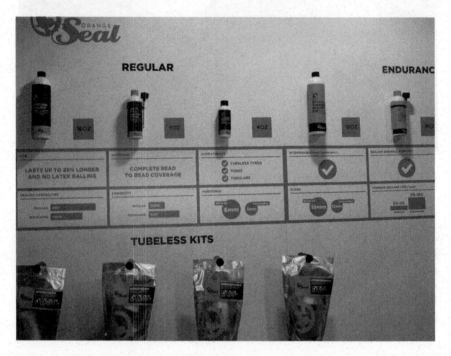

Figure 14.5 Display board for tubeless tyre kits.

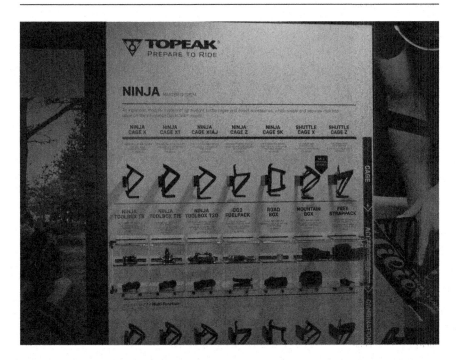

Figure 14.6 Accessory display board.

Race services

This can take two main forms: providing technical support for a particular team or providing what is called neutral services. Big race teams such as those with **Word Tour** status have their own full-time dedicated service team. The service team will service and repair the bicycles throughout the whole year ensuring that it is perfect for the rider from the early season training camps to the final end of season appearance events. This includes cleaning the bikes and transporting them between locations.

Lesser rated professional/semi-professional teams contract in service only for the events which they consider major to their race calendar. So, the service crew tends to work elsewhere during the rest of the year.

There are also neutral service crews that provide service to anybody, or team, at an event, this may be part of a publicity campaign to promote their products to riders and the public.

Specialist parts and services

With bicycles and their components becoming more complex there is an increasing need for specialist parts and services. The expense and complexity items such as electronic gears, hydraulic disc brakes and carbon fibre wheels require a large investment in both stock and special tools, and of course, the specially equipped van to transport them in.

The following figures 14.7, 14.8, 14.9, 14.10 were kindly provided by Frank Groom of Contraworx, www.contraworx.com

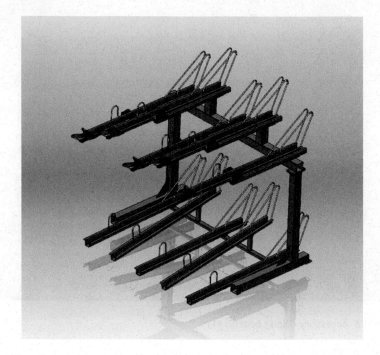

Figure 14.7 Double stacker.

Figure 14.8 Henley shelter.

Figure 14.9 Semi-vertical rack.

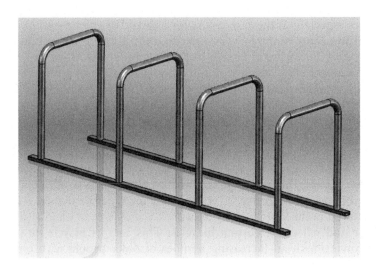

Figure 14.10 Sheffield hoop.

Mobile services: A number of small businesses are setting up to provide mobile services, typically visiting set locations at set times. This may be to a country town market, a busy shopping mall, or a parking lot in a small village. With the increasing costs of business rent and rates, the mobile business is a very attractive low-cost business concept.

Holidays: A growth area is the bicycling holiday market; this can take several forms, such as:

- Arrive and ride—bicycles provided, meet at airport, accommodation and meals provided.
- Own bicycle—transfer service and accommodation.
- Fully guided tours with **sag-wagon**—the following van for repairs or tired riders.
- Use of sat-nav which is pre-set to act as a guide.

Touring: Llike the holiday concept, it takes different forms; the main one is that a van, acting also as a sag-wagon, carries the luggage between the different night-stops. In other words, bicycle touring made easy.

Spinning: This is growing to be very popular. There appear to be two main types of spinners; those doing it to keep fit and maybe lose weight, others using it as winter, or bad weather training. This is very popular in areas of high traffic density, or where there are other reasons for not being able to get out on the road, including recuperating from illness.

Training camps: Usually operated in warm and sunny locations with high altitude steep hills. The lower air pressure at high altitude gives hypoxia—oxygen deficiency; this leads to the creation of more red blood cells and muscle tissue changes through a process known as the simulation of natural erythropoietin levels—commonly called EPO. Altitude training is normal for professional cyclists at the start of the season and before **World Tour** events. Many amateur riders use this too, especially if they are trying for a record or championship.

Transportation: Many bicyclists like to tour different parts of the country and do not want to ride to the area, nor wish to use a car. The transport of bicycles on trains and aeroplanes can be problematic. So, using specialist transport, taking care of the bicycle and luggage provides a good answer. This business is also growing in the motorcycle market, particularly for city-based riders who want to ride off-road in areas such as Scotland. There are a number of specialist bus services carrying bicycles and riders across popular European countries.

Bicycle Concierge: This concept is about bicycling for a very busy business person. Having the bicycle and kit ready when he/she gets off the aeroplane. This service is very popular for supercar owners—Ferrari waiting in the valet parking area.

Make and sell direct/custom bicycles: This area of bicycle business is growing with an increasing number of older people with high disposable income in most countries. Not about named brands of the bicycle; but about functionality.

Recycling—social enterprise: Also see Elephant Bike. This is also a hobby business for many individuals. By hobby business is meant one which covers expenses only, does not make a taxable profit. Repairing and restoring bicycles pays for the costs of owning one. A good way to start if still at school or college—the author did this as a school boy. It is possible to find old bicycles and components at re-cycling centres in most towns.

Essential service model: There are several essential bicycle parts which wear out or get damaged, then the owner is seeking replacements as cheap as possible and as quickly as possible. The motor trade calls this a distress purchase. A purchase we don't want to make, but have to make to remain on the road, that is a purchase because the owner is distressed. Rather like the motor trade, this includes tyres, tubes, batteries, bulbs and brake parts.

Hobby market/paraphernalia/fan-zone/collectables: Providing those bit and bobs that sit on our shelves and hang on our walls. The author collects items relating to The Clarion Cycling Club and Hill Special Cycles. He often finds himself bidding against friends who collect the same range of items.

Components/tools: Many bicycle components wear out, or the owner looks to up-grading them. There is a growing market in specialist and custom components, especially those made from carbon fibre, or billet aluminium, or billet titanium. Especially if it is an easy replacement procedure which can be carried out by the owner with a standard tool kit. Indeed, the market for specialist tool kits in carrying cases is growing too.

Cycle hire: A growing part of the holiday and leisure industry across the world. Flat seaside towns and country areas with un-made roads offer particularly good opportunities because riding is easy and safe. Charges are usually by the hour.

Accessories: Accessory sales depend on having an attractive item, at a good price and marketing it correctly—it's all about the unique selling point—USP. See Chapter 8 Add-ons and Kit.

Sportives: Often these are organised for charity, sometimes as commercial ventures. Etape type events are hugely supported—following the route of **World Tour** and **Monument** events.

TRADE ASSOCIATIONS

There are three well-known trade associations which offer services to the cycle industry in the UK, these are:

- Association of Cycle Traders
- The Bicycle Association UK

- Federation of Small Businesses—not specifically about bicycles, but gives a wide range of services and support to any small business

In the USA, there is the National Bicycle Dealers Association, other countries have similar associations. These are set up to give advice and guidance to businesses, and to lobby politically for the benefit of such businesses.
  Support from these associations variously includes:

- Technical training
- Reduced cost finance and banking
- Advertising materials
- E-commerce solutions
- Tax advice
- Legal advice
- Professional business promotion
- And many other services. As membership is not expensive, it is normal to be a member of more than one association

Chapter 15

Science terminology

Bicycles have now become highly technical pieces of equipment and are made to exact standards and tolerances. This is a long way from the bicycles made in the 19th Century, although then they were made to the standards and accuracies of that time. This section sets out to give the modern bicycle engineer, or technologist, an understanding of the terminology currently in use.

SI SYSTEM AND COMMON UNITS

SI stands for System International, a system of measurement units which was developed following World War II. There are several different systems in use throughout the World, but for your examinations with UK based examining bodies, SI only is used. It is worth noting that in countries such as Germany and Japan, they use SI, but with amendments and modifications. The Germans use DIN—Deutsch (Germany) Industrial Norm. The Americans use ANSI—American National Standards Institute, as well as SI. We'll also discuss some of the others as they are used in America too. The Imperial System—so called after the British Empire of the Victorian era.

Table 15.1 SI units

Quantity	Quantity symbol	Unit	Unit symbol
Length	L	Metre	M
Mass	M	Kilogram	kg
Time	T	Second	S
Electric current	I	Ampere	A
Temperature	T	Kelvin	K

Imperial System of Measurement for length uses inches, feet, yards and miles. For mass, it uses ounces, pounds and tons.

Table 15.2 is an **approximate** guide.

Table 15.2 Imperial/SI approximations

Quantity	Imperial	SI
Length	1 inch (In.)	25 mm
Length	1 foot (ft.)	300 mm
Length	1 yard (yd.)	900 mm
Length	39 inch	1 metre
Length	1 mile	1.6 kilometres
Mass	1 ounce	25 grams
Mass	1 pound	454 grams
Mass	2.2 pounds	1 kilogram (kg)
Mass	1 ton	1000 kilogram

Please note that the Glossary and List of Abbreviations at the back of the book gives more information about units and related terminology.

Table 15.3 Frame sizes, seat tube and top tube lengths

mm	Inches, to nearest half inch
50	19 ½
51	20
52	20 ½
53	21
54	21
55	21 ½
56	22
58	22 ½
59	23
60	23 ½
61	24
62	24 ½
63	25
64	25
65	25 ½

The engineering convention is to measure tube lengths centre to centre. That is the centre of one end lug to the centre of the other. However, in bicycle seat tubes, lots of makers measure them from the centre of the bottom bracket to the top of the seat tube. This may in fact include an extension above the top tube, especially with inclined top tubes.

Wheel sizes

Almost universally 700 mm in diameter, equivalents are 26-inch and 27-inch sizes. Some 29-inch wheels are in use for ATB/MTB and other off-road bicycles.

Stems sizes

The most common size is 105 mm, that is, 4 inch. This distance is measured centre to centre between the handle bar and the head set.

Crank sizes

The most common crank length is 170 mm, that is, 6 ¾ inch. This is measured between the centre of the bottom bracket and the centre of the pedal spindle.

Wheel base

Typically, this is 1 metre, which is just under 39 ½ inch.

Audax and sportive distances

If you wondered why Audax events and sportive events, which are not specific place to place, have odd number in either kilometres or miles, try converting from one unit to another.

Table 15.4 **Kilometres and miles**

Kilometres	Miles
50	31.25
100	62.5
150	93.75
160	100
200	125
240	150
320	200
400	250

DECIMALS AND ZEROS

It's very easy to make mistakes with decimal calculations and the use of zeros. As you have seen in Table 15.2, the standard or basic units are often too big or too small in value. So, there is a series of multiples and submultiples which are used to make calculations easier. For instance, kilo—meaning thousand—when added to metre gives kilometer, in other words, 1,000 metres. Going in the other direction, we use milli—meaning one thousandth—and when we are talking about the very low voltages in vehicle electronics—we say millivolts.

Table 15.5 Multiples and sub-multiples

Prefix	Symbol	Power	Number
giga	G	10^9	1 000 000 000
mega	M	10^6	1 000 000
kilo	K	10^3	1 000
hecto	h	10^2	100
deca	da	10^1	10
deci	d	10^{-1}	0.1
centi	c	10^{-2}	0.01
milli	m	10^{-3}	0.001
micro	μ	10^{-6}	0.000 001

ACCURATE MEASURING

Accurate measuring—called metrology in engineering terminology—is used extensively in bicycle manufacturing. Older bicycles were made with much larger tolerances than current ones. Tight accurate tolerances on lug joints, brake and gear hangers give a number of advantages, these are:

1. **Improved bicycle quality and appearance:** Enabling the bicycle to be sold for the highest possible price.
2. **Reduction in wind noise:** Giving better or more enjoyable ride.
3. **Improved aerodynamics:** Improving performance, particularly for time trials.
4. **Enabling more economical use of materials:** Reducing manufacturing costs.

Five pieces of equipment which are used in vehicle manufacture are starting to find their way into bicycle manufacture, these are currently scientific instruments and require skilled usage, but easy to use versions are now becoming available. Let's briefly discuss each one:

Co-ordinate measuring machine (CMM): When we are making a brake bracket or gear hanger to fit a frame, the first important measurements will be the mounting points. All other measurements will be taken from these points which we call co-ordinates—like the points on a map. A CMM measures distances from co-ordinates, even on the most irregular-shaped object such as a rear stay. It does this to an accuracy of one micron (1 µm, a millionth of a metre). This is necessary as often the first prototypes are made by hand from the contours of clay models—called a buck. Therefore, the exact measurements may not be known.

Granite block: This is what its name says, a gigantic block of granite on which a bicycle can be stood. This block of granite will have a mass greater than that of the bicycle and will be supported on many hydraulic columns, so that it can be kept perfectly level. Even the most level road surface will have a natural curvature—the curvature of the Earth. This granite block is made dead square. It enables dead accurate measurements to be taken of the bicycle—measurements which can be used during the construction process. Frame construction is usually undertaken on jigs—sets of metal rails which support the frame tubes, with attachments to position the fittings and hangers. The measurements from the granite block are used to inform the measurements for the production jig. For quality sampling, complete bicycles can be measured on the granite blocks.

Lasers: These can be used for positioning component mountings during construction. Having accurate frame measurements—using the CMM and position measurements from the granite block. It is possible to transfer this information to the frame jig to ensure that the tubes are correctly positioned before fixing takes place—either welding or bonding. This is done by setting up one or more lasers on the outer part of the jig that will shine a light, or several spots will converge together, when the part is accurately positioned. The tube can then be bonded or welded into place most accurately.

Scanners: These can be used for measuring an irregular shaped object, such as a handle bars; can be done without starting to mark the object, then the measurements may be transferable to metal to make another one. This has two main uses. One in manufacturing, the scanner can be used to take the profile from the design mock up—or first hand-made metal part, convert it into a CADCAM file for the manufacture of a press die. The other is to scan in the profile of vintage bicycle parts to aid manufacturing new ones.

Wind tunnel: It is increasingly used by racing cyclists to measure drag.

$$\text{The Drag Force} = \tfrac{1}{2}\, \text{air density} \times \text{velocity squared} \times \text{frontal area} \times \text{drag co-efficient}$$

This might seem a lot of variable, but it isn't really. Most wind tunnels will give some form of read out of force, or you attach streamers to actually see what is happening. You can then alter the wind speed by turning the fan motor up. Altering the rider position or component shape will change the frontal area and changing the shape, or material, will alter the drag co-efficient.

CAPACITY AND VOLUME

Liquids in the UK and Europe are sold in litres, this can be in parts of a litre or multiples of litres—for example, half-litre or may be 5 litres. In the USA, it may be sold in pints or gallons.

A litre is defined as the volume of 1000 cubic centimetres—1000 cc. Water has a mass of 1 kg per litre at a temperature of 4°C. This is often referred to as density. Oil and other lubricants are lighter than water and have a density of about 0.9 kg per litre.

Sometimes, the term relative density (rd) or specific gravity (sg) is used. Both sets of words mean the same thing. That is the density of the oil is compared to the density of water. So, the oil,—irrespective of its volume— if 1 litre weighs 0.9 kg, will have a rd of 0.9.

Still used by some companies, and more so in the USA are pints and gallons. Beer is also frequently sold in pints too. Be aware that English pints and gallons are different to American ones.

English pints are made up of 20 fluid ounces—a little obscure—that means the volume of 20 ounces of water at 17°C. An English pint of water weighs a pound and a quarter. An English gallon is 8 English pints weighing 10 pounds. It is defined as 4.54 litres.

American gallon is defined as 231 cubic inches, that is, 3.78 litres. It, in water, weighs 8.34 lb. So, it is considerably smaller. An American pint is 16 fluid ounces.

The volume changes with temperature—the volume of both oil and beer are legally measured in the UK at 16°C.

TEMPERATURE AND HEAT

These two scientific terms are often misused, so let's get them cleared up, so that you know what you are talking about when welding or brazing.

> **Temperature:** This is the hotness or coldness of an object. There are three temperature scales in use:
> **Celsius (C):** Also known as Centigrade because it has 100 degrees in it. It is related to the freezing and boiling point of water. Water freezes at 0°C and boils at 100°C.
> **Fahrenheit (F):** Water freezes at 32°F and boils at 212°F.
> **Kelvin (k):** Just uses the letter k—it is the absolute temperature scale. 0°C equals 273 k. 100°C equals 373 k. Absolute zero—lowest temperature achievable is 0 k which equals –273°C.
> **Heat:** This is the amount of energy used to raise the temperature something. Heat is a form of energy. It takes 4200 joules of heat energy to raise the temperature of one kilogramme of water 1°C. Water is said to have a specific heat of 4200 J/kg C.

When you are using a gas torch to warm up something, for instance a piece of metal which you wish to bend, you will note that it takes time. Different metals take different times, and larger pieces take longer than smaller one of the same metals. The longer time means that it is using more gas, this means more heat. As an example, typically 1 kg of propane will give 50 MJ.

FORCE AND PRESSURE

These two terms are also often confused, or misused, let's clarify them— it'll come in useful when you are pushing and bending, or straightening something.

Force

We often use this in calculations, but don't necessarily understand it. The unit of force is the newton (N), named after Sir Isaac Newton who first discovered it. He noticed that if anything was dropped, it would go to the ground. This is due to the force of gravity (G). The further an item drops the faster it goes—this is called acceleration due to gravity. The rate of acceleration on earth is typically 9.81 m/s/s. For terms of simple calculations, we often use 10 m/s/s as the equivalent of G:

$$\text{Force(N)} = \text{Mass(kg)} \times \text{Acceleration(G)}$$

Tech note

Mass is another name for weight when we are doing calculation with earth bound objects.

So, imagine an average person–mass 65 kg, stood on the roof of a car:

$$\text{Force of persons feet on car roof} = 65\,\text{kg} \times 10\,(\text{value of G})$$
$$= 650\,\text{N}$$

That is a static force. If we have a dynamic force—say swinging a hammer, then to get an approximate value, we call it 2 sigma; in other words, we double the total force. So, for a large sledge hammer of 5.5 kg (12 lb) with a moderate swing, we get 2 (5.5 × 10) = 110 N.

Pressure

Pressure is force divided by the cross-sectional area:

$$\text{Pressure}(\text{N/m}^2) = \text{Force(N)} / \text{Cross-sectional area } (\text{m}^2)$$

To avoid confusion with other units, the term Pascal (Pa) is used for 1 N/m^2 pressure.

The pressure of 1 Pa is very low—imagine an apple on your desktop that's about 1 Pa. So, we tend to use the term bar—this is equivalent to normal barometric pressure. 1 bar equals to 101.3 KPa. The air compressor which provides air pressure for your power tools generates about 10 bar in pressure. Bicycle tyres are typically inflated to between 4.5 bar on a mountain bike and 7.5 bar on a racing bike.

AMPS, VOLTS, OHMS, WATTS AND KIRCHHOFF

Electricity is easy to understand, if you get the basic terms clear. Although you can't see electricity, it behaves in a similar way to water. Providing that the plumbing in your workshop is connected to the mains supply, when you turn the tap on water will flow out. Water comes out under pressure—a pressure from the main supply. The consumer standard for water pressure in the UK is enough pressure to fill a 4.5 litre bucket in 30 seconds; typically, about 2 bar.

When we talk about electricity, the switch replaces the tap, the voltage (V) replaces the pressure and the amps (I) replaces the bucket full. So, the voltage needs to be high enough to force enough amps through to light up your lamp, or power the items in a circuit provide a resistance (R), like the tap, they can slow or stop the flow of electricity.

SAFETY NOTE

Electricity can kill you

Please be aware that any voltage or amperage of electricity can kill you. It can also give you a non-lethal shock, which can make you flitch or jump and cause personal injury by hitting a rotating part, or a hot part.

Ohm's Law

The relationship between amps (I) and volts (V) is given below, which will give a value for the resistance measured in ohms (R).

$I = V/R$

Watt (W)

This is a measure of power. If you look at light bulbs, they will usually have their power rating on them, the same applies to electric motors and heating elements. It is normal to state the power and the voltage on these electrical items. With LED lights, the equivalent wattage is often given in two ways. For instance, some mains LED lights are rated as 9 W = 100 W, that is, they consume 9 W but give the equivalent light of a 100 W tungsten bulb—the older type light bulb. They still work off a 230 V mains power supply.

$$Watts(W) = Volts(V) \times Amps(I)$$

Tech note

I = Current in Amps.

  A is often used in a colloquial, or less formal way, also meaning Amps.

If you have the wattage and the voltage, you can work out the current consumption in amps:

$$I = W/V$$

Knowing the current consumption is useful when you are fault finding, or wiring up a new component. Fault finding you can use an induction ammeter to see the actual current flowing. When fitting a component, it allows you to choose the correct cable size.

Kirchhoff

At any junction in an electrical circuit, the current flowing into the junction will equal the current flowing out. This gives you more information when testing a circuit—finding where the current is flowing to.

SAFETY NOTES

Always isolate the circuit concerned.

Disconnect the battery where appropriate.

FRICTION

Skidding: Happens when the friction between the tyres and the road is not sufficient to keep the bicycle on a course. Friction is the ratio between the force acting downwards on the tyre (weight—W) and the force (F) needed to slide the tyre over the road. Not rolling the tyre, but making it skid. On a normal road with good tyre, this ratio, μ (Greek letter called Mu) expressed as a decimal fraction, is about 0.8. If the road is covered in ice or wet leaves, the ratio can be as low as 0.01; in other words, it can be pushed along with its brakes on.

μ = Force / Weight

SOME COMMON LAWS OF MECHANICAL ENGINEERING

Newton's Laws are about force and acceleration they are numbered:
First Law: A body—bicycle, piece of metal, etc.—will either stay where it is, or continue moving uniformly unless another force is applied to it.
Second Law: The force on a body is equal to its mass multiplied by its rate of acceleration. Usually expressed in the form Force (F) = Mass (m) × Acceleration (a).
Third Law: When a force is applied to a second body, the second body will be exerting a force backwards. Sometimes said as, to each action there is an equal and opposite reaction.
Hooke's Law: It is about how metal reacts to force. Metal stretches by an amount (X) proportionally to the force (F) applied to it until it gets to its elastic limit when it breaks. The amount of stretch depends on the type of metal, the metal type is given as constant (k)

$$F = kX$$

IMPACT AND MOMENTUM

Momentum: When a bicycle is travelling along a road, it processes momentum—Newton's First Law. That momentum (p) is the product of its mass (m) and velocity (v) — that is speed combined with direction.

$$p = mv$$

The heavier the vehicle and the faster the speed, the greater is the momentum. This is why we have speed limits.
Kinetic energy: We know that the moving vehicle possess momentum and again applying Newton's Laws—this time the Second Law, we have to apply a force to stop.

Chapter 16

Service and repair

INTRODUCTION TO COSTINGS AND THE BUSINESS MODEL

Setting up a bicycle business either falls into manufacturing or retailing. This section is to help the new comer wishing to set-up in the retail sector.

Tech note

See also the Chapter 14 on the Bicycle Industry for manufacturing and other business ideas, and the Chapter 12 on Data for business calculations and forms.

The bicycle retail sector is changing very quickly; however, business principles remain very much as they have for thousands of years. In the bicycle retail sector, there are two main income streams, these are:

1. Buying bicycles and bicycle parts and accessories and selling them for a higher price—the difference between the wholesale price and the retail price is called mark-up, typically the amount of mark-up is between 25% and 50% of the retail price—not including the VAT or other taxes which are added on to the final retail price.
2. Selling labour or time, that is, time used for the servicing and repair of bicycles and bicycle components. The time may be provided by the shop owner, or by employed staff. This will be subjected to VAT or other taxes, which are added to the final bill.

Tech note

All businesses carry some form of financial risk—the biggest is buying stock and offering services which nobody wishes to buy—at which point you lose money.

Successful retailing of bicycles and parts requires two important ingredients, these are buying saleable items at a price which you can sell them to make a profit, and having a variety of ways of selling them—typically in-store, mail-order and specialist shows.

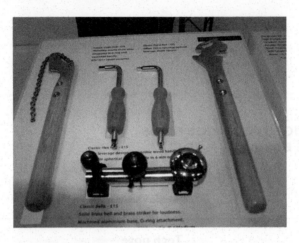

Figure 16.1 **Range of workshop tools 1.**

Figure 16.2 **Range of workshop tools 2.**

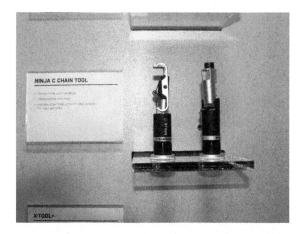

Figure 16.3 Tools to fit in ends of handle bars.

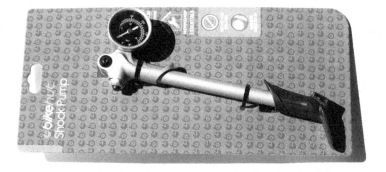

Figure 16.4 Pneumatic fork pump.

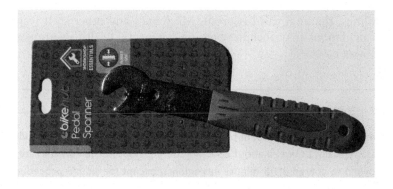

Figure 16.5 Chain tool.

Figure 16.6 Range of tools.

Successful selling of time, or offering services, requires having the necessary tools and skills to provide the services that customers want to buy.

Generally, the more expensive an item is the greater will be the profit, but you may sell less of them because of the high price point. With service skills, specialist high level skills generate a higher or premium rate, for the same amount of time.

Rule of thumb on bicycle mechanics wages is that they will be paid one-third of the charge-out rate. Another third goes to cover overheads—rent and utilities, the final third is profit. Mechanics and other staff wages in these calculations may include: pension payments, health insurance (NI) and holiday pay.

Tech note

Charge-out rate is the amount per hour that the customer pays, not including taxes. Typically, between about £50 and £80 per hour at the current rate. It is normal to charge in 15 minute increments, except for simple procedures where you will have menu pricing.

TYPICAL SERVICE/REPAIR PROCEDURES

Dr Bike

This is a concept used by a number of cycling organizations in an attempt to make cycling safer. Sometimes it is combined by Police post-code marking bicycles to reduce crime. A number of bicycle repair companies also run Dr Bike sessions, besides raising safety awareness, it can also generate extra business. If a session is run correctly, it can be a good public relations exercise, give safety awareness, spread bicycle knowledge and help to generate business. Visits to school and shopping malls appear to be the most successful way of running these events.

Check list: This check list is broken down to cover the main areas of a Dr Bike session, answers are yes or no. If no, the point would be further investigated and the owner advised appropriately. A Dr. Bike check should take about ½ hour.

Table 16.1 Dr Bike checklist

Area	Specific	Y/N
Front wheel	True (not buckled). No broken/missing spokes. Good rim.	
Front tyre	Good tread. No splits, cracks or holes. Pumped hard. Valve straight.	
Front hub	No wobble. Turns smoothly. Wheel securely fixed.	
Front mudguard	Firmly fixed. No sharp mudguard stays.	
Front brake blocks	Correctly fitted. Not worn away.	
Front brake	Firmly fixed. Correctly adjusted.	
Front brake lever	Comfortable position. Firmly fixed. Cable not frayed.	
Headset/steering	No wobble. Correctly adjusted.	
Handlebars	Not distorted. Ends protected.	
Front forks	Appear true and undamaged.	
Frame	Appears true and undamaged.	
Rear brake lever	Comfortable position. Firmly fixed. Cable not frayed.	
Rear brake	Firmly fixed. Correctly adjusted.	
Rear brake blocks	Correctly fitted and aligned. Not worn away.	
Rear mudguard	Safely fixed. No sharp mudguard stays.	
Rear tyre	Good tread. No splits, cracks or holes. Pumped hard. Valve straight.	
Rear wheel	True. No broken/missing spokes. Good rim.	
Rear hub	No wobble. Turns smoothly. Wheel securely fixed.	
Bottom bracket	No wobble. Lock ring tight. Sufficiently lubricated.	
Pedal cranks	Straight.	
Pedals	Complete. Turning freely. Not bent.	
Chainwheel	Not bent. Teeth not worn.	

(*Continued*)

Table 16.1 Dr Bike checklist (Continued)

Area	Specific	Y/N
Chainguard	Firmly fixed. Not bent.	
Chain	Not too worn. Not slack. Lightly oiled not rusty.	
Gears	Properly adjusted. Lubricated sufficiently.	
Saddle	Safely fixed. Straight, comfortable height (unless BMX).	
Rack/carrier/bags etc.	Firmly secured.	
Front lamp (if carried)	White. Firmly fixed. Good light to front.	
Rear lamp (if carried)	Red. Firmly fixed. Visible to rear.	
Reflectors	Clean and secure.	

Preventative maintenance

Like your car, servicing the bicycle will prevent breakdowns and accidents. The usual method is like car servicing too, what is called menu servicing, a number of packages offered for fixed prices. These usually take the form of basic, middle and best—in the example labelled bronze, silver and gold. The pricing is based on the time and effort involved for each, typically this will be bronze ½ hour, silver 1 hour, gold 2 hours. Parts will be charged at normal retail price, so adding to the profit margin for the job. For example, if a chain is replaced on the bronze service, this will take less time than cleaning the original one and give a profit on the sale of the new chain. Both customer and repair shop win, everybody is happy.

Table 16.2 Service menu

No.	Tasks	Bronze	Silver	Gold
1.	Safety check of complete bicycle	X	X	X
2.	Brake balancing and servicing	X	X	X
3.	Clean or replace chain	X	X	X
4.	Fit new tyres/tubes if needed	X	X	X
5.	Service or replace brake cables		X	X
6.	Service or replace gear cables		X	X
7.	Check and fine tune gear change		X	X
8.	Service or replace cassette/sprocket		X	X
9.	Clean and check/replace chainset		X	X
10.	Clean and check/replace pedals		X	X
11.	Inspect/replace bottom bracket			X
12.	Inspect and adjust headset			X
13.	Check and true wheels if needed			X
14.	Service front hub			X
15.	Service rear hub			X
16.	Replace bar tape/grips			X
17.	Waterless clean			X

LEGAL REQUIREMENTS

These regulations are applicable to the UK, although in other countries the rules are similar.

Road Vehicle Lighting Regulations (RVLR)

These are regularly updated, the most recent is 2009. The main points are:

- Lights and reflectors are required in bicycles between sunset and sun rise.
- Lights are not required if the bicycle is stationary or being pushed along the road side.
- Lights and reflectors must be clean and in proper working order.
- The white front light must be affixed to the bicycle, either centrally or to the off-side (right), it must have a steady minimum output of 4 candelas, that's about 50 lumens. The maximum height is 1,500 mm from the ground. It must conform to BS6102/3or equivalent.
- The red rear light must be affixed to the bicycle, either centrally or to the off-side (right), between 350 mm and 1,500 mm from the ground, at or near the rear, aligned towards and visible from behind. It must have a steady minimum output of 4 candelas, that's about 50 lumens. It must be marked as conforming to BS3648, or BS6102/3, or an equivalent EC standard.

Flashing bicycle lights

Current practise is for the use of flashing bicycle lamps both front and rear. The standards do not cater for this type of light, however, making the legal situation complex. The 2005 RVLR amendment means that it is now legal to have a flashing light on a pedal cycle, provided it flashes between 60 and 240 times per minute (1–4Hz). With the absence of standards, check that it is made by a well-known reputable manufacturer.

Rear reflector

One rear reflector is required, coloured red, marked BS6102/2 (or equivalent), positioned centrally or offside (on the right-hand side of the bike), between 250 mm and 900 mm from the ground, at or near the rear, aligned towards and visible from behind.

Helmet mounted lights

Simply these do NOT comply with the lighting laws. The lights must be attached to the bicycle.

Brakes

Bicycles must be fitted with two independent braking systems. On a fixed gear machine, the rear brake may take the form of pedal resistance, so only a front brake is fitted.

SECURITY

Bike register

It is a police approved national register of bicycles. It is free to join. Bicycles can be post code marked and listed on the register with the frame number.

Figure 16.7 University bike repair stand—Cambridge School of Art 1.

Figure 16.8 University bike repair stand—Cambridge School of Art 2.

Figure 16.9 University bike repair stand—Cambridge School of Art 3.

ASSOCIATION OF CYCLE TRADERS

The Association of Cycle Traders (ACT) is the largest cycle trade association in the UK and has been promoting cycling for over 100 years. The ACT is a membership organization representing the interests of over 4000 businesses involved in the cycle industry through promotion, business support and skills development. If you're not already involved you can even become part of the community for free.

TRAINING AND QUALIFICATIONS

There is a small number of education and training organizations which offer specialist training for what is commonly referred to has *cycle mechanics*. They usually offer some form of certification, this may take the form of an assessed certification from one of the awarding organizations—called AOs or just a certificate of completion by the trainer.

The two best known AOs in bicycle training are The IMI and C&G. The content of the qualifications offered by these two organizations is based around the Occupational Standards, which have been developed by a Government organization called The Institute for Apprenticeships and Technical Education, known as The Institute. The Institute is an executive non-departmental public body, sponsored by the Department for Education.

Tech note

The apprenticeship standard for cycle mechanics can be found in Appendix 2.

In the field of cycle mechanics, there are three levels of vocationally related qualifications—VRQs, these are:

- **Level 1:** Sub-GCSE level equivalent—covers basic service and repair procedures.
- **Level 2:** GCSE level equivalent—covers the knowledge and skills of a professional repair mechanic.
- **Level 3:** A level equivalent—covers advanced skills as appropriate to a technician.

Because of its specialist nature, it is normal to start at Level 1 no matter what your previous level of education is.

In addition to completing either, or both, an apprenticeship and/or a VRQ, it is normal to complete specialist training courses for topics like gear repairing and wheel building. If you are interested in further training, then the first contact should be to The Association of Cycle Traders, they strongly support training within the industry.

SCREW THREADS

To any engineer screw threads pose numerous pitfalls. In the early days of bicycles, each manufacturer made screw threads on an individual basis, remember that we have had them a long time. Archimedes used screw threads in the 3rd Century BC. In 1841, Joseph Whitworth proposed a set of standards for threads, the problem for bicycle manufacturers is that the Whitworth tread is too course for small bicycle components. At that time, the Birmingham Small Arms Company (BSA) was being formed; this was a group of gun manufacturers who eventually went on to make bicycles, motorcycles and cars. They adapted the gun threads for bicycle use. This led to threads known as cycle threads, but there are two variations of this, these are the Cycle Engineers' Institute thread which was superseded by the British Cycle Standards (BSC) thread. In current terminology, cycle threads tend to be called BSA threads giving reference to their heritage; hence we have BSA bottom brackets. Of course, European and Asian manufacturers use their own variations of this, which may look similar until measured in detail.

CYCLE ENGINEERS' INSTITUTE THREADS

Imperial Wire gauge	Diameter Inches	Threads per inch	Tap Drill Size
17	0.056	62	61
16	0.064	62	56
15	0.072	62	54
14	0.080	62	1/16
13	0.092	62	49
12	0.104	44	47
—	1/8	40	2.5 mm
	0.154	40	30
	5/32	32	
	0.175	32	27
	3/16	32	23
	7/32	26	
	¼	26	4
	17/64	26	
	9/32	26	
	5/16	26	1
	3/8	26	8.5 mm
	7/16	26	
	0.4724 (12 mm)	26	

(Continued)

Imperial Wire gauge	Diameter Inches	Threads per inch	Tap Drill Size
	½	26	
	0.5512 (14 mm)	26	
	9/16	26 +20?	13 mm
	5/8		
	0.9675	30	
	I	26	24.5 mm
	1.29	24LH	1¼
	1.37	24	1 21/64
	1 7/16	24LH	35.5 mm
	1½	24	37 mm

BRITISH CYCLE STANDARDS (BSC) THREADS

These have a symmetrical 60° thread angle, with rounded roots and crests having a radius of 0.166 times the pitch. Actual depth, 0.5327 times the pitch. The smaller sizes are for spokes.

Imperial Wire gauge	Diameter Inches	Threads per inch	Tap Drill Size
17	0.056	62	61
16	0.064	62	56
15	0.072	62	54
14	0.080	62	1/16 inch
13	0.920	56	49
12	0.104	44	47
—	0.125	40	2.5 mm
	0.154	40	30
	0.175	32	27
	0.1875	32	23
	0.250	26	4
	0.266	26	I
	0.281	26	C
	0.3125	26	J
	0.375	26	8.5 mm
	0.5625	20	13 mm
	1.000	26	24.5 mm
	1.290	24	1¼ inch
	1.370	24	1 21/64 inch
	1.4375	24	35.5 mm
	1.500	24	37 mm

BRITISH STANDARD WHITWORTH CYCLE THREADS

Also, you will find British Standard Whitworth special cycle threads, these are similar to the BSA type cycle threads; but the thread angle is 55° whereas the BSA angle is 60°

Major diameter Inches	Threads per inch
7/16	20
½	20
9/16	20
5/8	20
11/16	20
¾	20

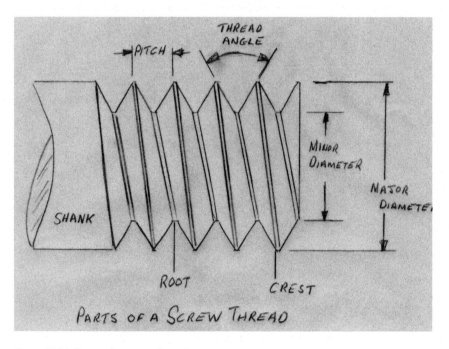

Figure 16.10 **Parts of a screw thread.**

Figure 16.11 Pedal spanner.

Glossary

This section defines a number of the words and phrases used in bicycle engineering and bicycle usage, including some of the specialist racer and enthusiast vocabulary and jargon.

'Esses' one bend followed by another
'O' rings rubber sealing rings
10/25/50/100 Time trial race lasting 10 or 25 or 50 or 100 miles, the winner is the one doing it in the shortest time
12-hour/24-hour Time trial race lasting 12 or 24 hours, the winner is the one covering the greatest distance
A head a system of clamping the stem to the outside of the fork
Acceleration rate of increase of velocity
Accessories anything added which is not standard on a bicycle
Add-ons something added after bicycle is made
Adhesion how the bicycle holds the road
Alignment position of one item against another
Allen key an hexagonal shaped bar-like tool
Allen screw a screw with a countersunk hexagonal head
Alloy mixture of two or more materials; may refer to aluminium alloy, of an alloy of steel and another metal such as chromium
Ally slang for aluminium alloy
Atom single particle of an element
Ball bearing bearings which are ball shaped
Bar end end of the handle bar, this should be plugged for safety
Barrel a hollow cylinder
Bearing any surface or joint which allows movement
Bench working surface; also flow bench and test bench
Beta version test version of software, or product
Bidon cyclists term for water/drinks bottle
Bonk pain in abdomen when short of food or suddenly riding too hard
Bore internal diameter of cylinder barrel

Bottom bracket the bearing assemble that supports the cranks and pedals

Bracket height height of the bottom bracket above the road

Brooklands first purpose built racing circuit at Weybridge in Surrey with banking and bridge

Carbon fibre like glass fibre but used with very strong carbon-based material

Chain stays tubes between rear axle and bottom bracket

Chicane sharp pair of bends—often in the middle of a straight

Chocks tapered block put each side of wheel to stop the bicycle rolling

Circuit race circuit

Clerk of Course most senior officer at a bicycle racing event—person whose decision is final, although there may be a later appeal to the various governing bodies

Cockpit area around handle bars

Code reader reads fault codes in the ECU of the particular system

Composite material made in two or more layers—usually refers to carbon fibre, may include an honey comb layer

Condensation change from gas to liquid

Contraction decreases in size

Corrosion there are many different types of corrosion; oxidation or rusting are the most obvious

Cotter pin a tapered pin which attaches the crank to the bottom bracket on some bicycles

Crank the rotating part between the bottom bracket and the pedal

Cushion section of seat to sit on

Dash board instrument panel

Density relative density, also called specific gravity

Derailleur a type of gear-changing mechanism that takes the chain from one sprocket to another

Diagnostic equipment connected to the system to find faults

Dive bicycle goes down at front under heavy braking, or rider is thrown off

Down tube tube between the bottom bracket and the head tube

Drag strip flat and smooth section of road—fast racing section of road

Drop-out components at the end of the seat stays and chain stays to locate axle

Epoxy resin material used with glass fibre materials

Evaporation change from liquid to gas

Event organizer person who organizes the race or other event

Expansion increase in size

Fast back long sloping rear forks

Fender USA for mudguard

Flag marshal marshal with a flag

Flag chequered flag, black flag, red flag and other colours used for different purposes

Foam material used to make seats and other items

Force mass multiplied by acceleration

Fork crown piece at top of forks

Fork ends pieces at the ends of the fork blades to attach the front axle

Fork component that holds the front wheel

Friction resistance of one material to slide over another

Frontal area (projected) area of front of cycle and rider

Gelcoat a resin applied when glass fibre parts are being made—it gives the smooth shiny finish

Glass fibre light weight mixture of glass material and resin to make vehicle body

Handle bar part held by rider's hands

Head set bearing mechanism allowing fork to turn inside head tube

Head tube tube between the top tube and the down tube

Heat a form of energy—hotness

Hill climb individually timed event climbing a hill

HPV human powered vehicle—non-standard bicycle such as stream-line one

Hub components that form the center of the wheel

Inboard something mounted on the inside of the drive shafts such as inboard brakes; usually lowers un-sprung weight

Inertia resistance to change of state of motion—see Newton's Laws, inertia of motion and inertia of rest

Kevlar fibre super strong material, often used as a composite with carbon

Le Tour common name for The Tour de France

Machine the bicycle

Marshal person who helps to control an event

Mass molecular size, for most purpose the same as weight

Metal fatigue metal is worn out

Molecule smallest particle of a material

Mono-fork fork with only one blade

Monument important race in international calendar

Monza World's second banked race track in Italy, copy of Brooklands

MTB mountain bike—bike for off-road use

Musette cotton shoulder bag used to feed riders on long distance races

Newton's Laws First Law—A body continues to maintain its state of rest or of uniform motion unless acted upon by an external unbalanced force; Second Law—The force on an object is equal to the mass of the object multiplied by its acceleration ($F = Ma$) and Third law—To every action there is an equal and opposite reaction

Nose cone detachable front body section covering front of a HPV—may include foam filler for impact protection

Off-roader bicycle for going off-road, or off-road event

Original finish original paint work, usually with reference to historic or vintage bicycles

Outboard something mounted on the outside of the drive shafts such as brakes, usually increases un-sprung weight

Oxidation material attacked oxygen from the atmosphere; aluminium turns into a white powdery finish

Paddock where teams and bicycles are based when not racing

Parent metal main metal in an item

Pedal the rotating component on the crank that the rider's foot turns

Pot another name for cylinder

Power work done per unit time, HP, BHP, CV, PS, kW

Prepping preparing the bicycle for an event

Prototype first one made before full production

Quick release a mechanism that allows a component to be removed quickly

Quill a feather-shaped cut; used on handle bar stems to grip the inside of the fork

Regs racing regulations

Ride height height of bicycle seat off the road, usually measured from road to the top of the saddle

Rings piston rings

Rust oxidation of iron or steel—goes to reddish colour

Screw thread helix cut on bolts and screws

Scrutineer person who checks that a bicycle compiles with the racing regulations, usually when scrutineered the bicycle has a tag or sticker attached

Seat stays thin tubes between rear axle and top of seat tube

Seat tube tube between the bottom bracket and saddle

Skewer thin rod inside axle to operate quick release

Skid bicycle goes sideways—without road wheels turning

Speed event any event where bicycles run individually against the clock

Spine backbone like structure

Sprint accelerating quickly for finish line

Sprinting individually timed event starting from rest over a fixed distance

Sprung weight weight below suspension spring

Squat bicycle or rider goes down at the back under heavy acceleration

Squeal high pitch noise—usually from brakes

Stage event when the event is broken into a number of individually timed stages, the bicycles start the stage at pre-set intervals (typically 2 min)

Stall involuntary stopping of bicycle and rider—usually due to fatigue

Start ramp downward incline used at the start of major time trial event

Steward a senior officer in the organization of the bicycle event

Straw bales straw bales on side of the track for a soft cushion in case of an off

Stress force divided by cross-sectional area

Stripping pulling apart

Stroke distance piston moved between TDC and BDC

Swage line raised design line on metal panel

Swage raised section of metal panel

TDC top dead centre

Team bus bus used to transport the team of riders and officials to events

Temperature degree of hotness or coldness of a body

Test bench test equipment mounted on a base unit

Test hill a hill of which the gradient increases as the top approaches, originally the test was who got the furthest up the hill

Top tube tube between the head tube and the down tube

Torque turning moment about a point (Torque = Force × Radius)

Track stand standing on the bicycle with feet on pedals

Transporter vehicle to transport bicycles and team equipment to events

Tub tubular tyre

Tyre wall wall on side of track built from tyres—giving a soft cushion in case of an off

Un-sprung weight weight below suspension spring

Velocity vector quality of change of position, for most purposes the same as speed

World Tour international stage race series

METALS

Aluminium is extracted from bauxite ore electrically—1 Kg of ore produces about 250 gm of aluminium. Production of aluminium is very expensive as it uses lot of electricity. Pure aluminium is too weak for most purposes, so it is alloyed with other metals such as copper, manganese, silicone, magnesium, tin and zinc.

Chromium small percentages are added to steel to give corrosion resistance; these are often referred to as stainless steel.

Cobalt used to add hardness to steel and enhance magnetic properties of steel. Steel with cobalt and chromium are sometimes referred to as stellites.

Copper an excellent conductor of both heat and electricity. When it oxidizes it turns green. It work-hardens easily. It is often alloyed to form brass, bronze and cupro-nickel and nickel silvers.

Germanium used as a semi-conductor.

Gold best electrical and thermal conductor, very ductile and readily cold worked.

Iron the basis of all steels, alloyed with other metals and defined by its carbon content.

Lead very malleable, good electrical conductor.

Magnesium a very low-density material, used to alloy with aluminium to improve machinability. It burns ferociously and reacts with steam.

Molybdenum it has a high density, and is a good electrical and thermal conductor. It is alloyed with steel to improve hardness and corrosion resistance.

Nickel alloyed with steel to increase corrosion resistance and strength at high temperature.

Niobium alloyed with steel because of its high melting point and high resistance to corrosion.

Palladium alloyed with gold, silver and copper for its high resistance to corrosion. Often, this is combined with iridium and rhodium in technical instruments.

Silicone used as a semiconductor.

Silver good thermal and electrical conductor.

Tantalum acid-resistant and high melting point, often alloyed with tungsten.

Tin low strength metal with high corrosion resistance. Alloyed with lead and antimony for bearing surfaces.

Titanium very high strength and very low density with good corrosion resistance.

Tungsten high density and the highest melting point of all metals. Used as an alloy with steel.

Zinc high corrosion resistance, used to coat steel to prevent rusting—the process is called galvanizing.

Index